CATALYTIC HYDROPROCESSING OF PETROLEUM AND DISTILLATES

CATALYTIC HYDROPROCESSING OF PETROLEUM AND DISTILLATES

Based on the
Proceedings of the
AIChE Spring National Meeting,
Houston, Texas
March 28 – April 1, 1993

edited by

Michael C. Oballa

Novacor Research and Technology Corporation
Calgary, Alberta, Canada

Stuart S. Shih

Mobil Research and Development Corporation
Paulsboro, New Jersey

CRC Press
Taylor & Francis Group
Boca Raton London New York

CRC Press is an imprint of the
Taylor & Francis Group, an **informa** business

CRC Press
Taylor & Francis Group
6000 Broken Sound Parkway NW, Suite 300
Boca Raton, FL 33487-2742

First issued in paperback 2019

ISBN-13: 978-0-367-40202-0

Library of Congress Cataloging-in-Publication Data

National Meeting of AIChE (1993: Houston, Tex.)
      Catalytic hydroprocessing of petroleum and distillates :
proceedings of the AIChE Spring National Meeting, Houston, Texas,
March 28–April 1, 1993 / edited by Michael C. Oballa and Stuart S.
Shih.
      p. cm. -- (Chemical industries; v. 58)
      Includes bibliographical references and indexes.

      1. Catalytic cracking--Congresses. 2. Hydrotreating catalysts-
-Congresses.   I. Oballa, Michael C. II. Shih, Stuart S. III. Title.
IV. Series.
TP690.4.N38 1993
665.5'33--dc20   94-12079
                                                                         CIP

**Visit the Taylor & Francis Web site at
http://www.taylorandfrancis.com**

**and the CRC Press Web site at
http://www.crcpress.com**

Preface

The era of refineries that process only light oils is almost over. There is a strong push for the processing of heavy oils, bitumen and/or residue, which in commercial jargon are combined and termed "bottom-of-the-barrel." The processing of "bottom-of-the-barrel" carries with it some problems. These are connected with obtaining state-of-the-art technologies at reasonable capital and operating costs to the refiner. Then there are problems associated with choosing the best catalyst—one specially designed to lower considerably the high content of heteroatoms (S, N, O) and metals (V, Ni, Fe) in the "bottom-of-the-barrel." The effective life of such catalysts becomes a factor to be reckoned with, especially when the economics of the various processes is being looked at. Recent regulations on the content of aromatics and sulfur in diesel fuels, as well as environmental concerns on SO_x and NO_x emissions from mobile and stationary combustion facilities, have made it imperative for engineers and scientists to think of ways to economically process middle distillates to meet new product specifications or more stringent emission regulations.

To address the above consideration, we organized a symposium at the AIChE Spring National Meeting in Houston, March 28–April 1, 1993. Engineers and scientists working in the processing of petroleum and distillates from different parts of the world presented papers covering different facets of residue upgrading and distillate hydrotreating. We thank the Fuels and Petrochemicals Division of the American Institute of Chemical Engineers for sponsoring this symposium. This book is a compilation of most of the papers presented in the five sessions of the symposium. We have broadly classified the papers in terms of content into the following four categories:

a) Catalyst Deactivation
b) Upgrading of Heavy Oils and Residue
c) Hydrotreating of Distillates
d) General Papers

We would like to thank all the people who contributed to make the publication of this book possible, including the authors, speakers and various organizations that provided the necessary support. We give special thanks to the authors for preparing the manuscripts in the requested form and of such high quality that little editorial work was necessary.

We also express our appreciation to Professor Carle H. Bartholomew, who, as an invited guest speaker at the symposium, agreed to prepare a review paper on Catalyst Deactivation for this publication. Finally, the secretarial help of Ms. Karen Armellino and Ms. Kerri Dauphinee of Novacor Research & Technology Corporation is gratefully acknowledged.

Michael C. Oballa
Stuart S. Shih

Contents

Contents

GENERAL PAPERS

Contributors

M. Absi-Halabi   Petroleum Technology Department, Kuwait Institute for Scientific Research, Safat, Kuwait

H. Al-Zaid   Petroleum Technology Department, Kuwait Institute for Scientific Research, Safat, Kuwait

C. H. Bartholomew   Department of Chemical Engineering, Brigham Young University, Provo, Utah

Chakib Bennouna   Laboratoire de Chimie Organique Appliquée, Faculté des Sciences, Université Cadi Ayyad, Marrakech, Morocco

Abdennaji Benyamna   Laboratoire de Chimie Organique Appliquée, Faculté des Sciences, Université Cadi Ayyad, Marrakech, Morocco

J. S. Brinen   CYTEC Research and Development, CYTEC Industries, A Business Unit of American Cyanamid Company, Stamford, Connecticut

Ph. Caillette   Institut Français du Pétrole, Rueil-Malmaison, France

C. N. Campbell   Texaco Research and Development, Port Arthur, Texas

J. D. Carruthers CYTEC Research and Development, CYTEC Industries, A Business Unit of American Cyanamid Company, Stamford, Connecticut

G. A. Clausen   Texaco Research and Development, Port Arthur, Texas

Barry H. Cooper   Haldor Topsøe A/S, Nymøllevej, Lyngby, Denmark

E. P. Dai   Texaco Research and Development, Port Arthur, Texas

J.-C. Duchet   Laboratoire Catalyse et Spectrochimie, Université de Caen, France

Patrick Geneste   Laboratoire de Chimie Organique Physique et Cinétique Chimique Appliquées, Ecole Nationale Supérieure de Chimie, Montpellier, France

G. Germaine   Shell Recherche SA, CRGC, Grand-Couronne, France

M. Gjers   Shell Raffinaderi AB, Gothenburg, Sweden

S. Greenhouse   CYTEC Research and Development, CYTEC Industries, A Business Unit of American Cyanamid Company, Stamford, Connecticut

Teh C. Ho   Corporate Research Laboratories, Exxon Research and Engineering Company, Annandale, New Jersey

J. R. Huang   Texaco Research and Development, Port Arthur, Texas

S. Kasztelan   Institut Français du Pétrole, Rueil-Malmaison, France

Z. Khan   Petroleum Technology Department, Kuwait Institute for Scientific Research, Safat, Kuwait

M. T. Klein   Center for Catalytic Science and Technology, Department of Chemical Engineering, University of Delaware, Newark, Delaware

Peter Kokayeff   Unocal Science and Technology Division, Brea, California

D. A. Komar   CYTEC Research and Development, CYTEC Industries, A Business Unit of American Cyanamid Company, Stamford, Connecticut

S. C. Korré   Center for Catalytic Science and Technology, Department of Chemical Engineering, University of Delaware, Newark, Delaware

Klaus Kretschmar VEBA OEL Technologie und Automatisierung GmbH, Gelsenkirchen, Germany

Andrzej Krzywicki Novacor Research and Technology Corporation, Calgary, Alberta, Canada

R. N. Landau * Center for Catalytic Science and Technology, Department of Chemical Engineering, University of Delaware, Newark, Delaware

J. Leglise Laboratoire Catalyse et Spectrochimie, Université de Caen, France

Leszek Lewkowicz Alberta Research Council, Edmonton, Alberta, Canada

J. P. Lucien Companie Rhenane de Raffinage Reichstett, Reichstett Vandenheim, France

N. Marchal Institut Français du Pétrole, Rueil-Malmaison, France

V. K. Mathur Department of Chemical Engineering, University of New Hampshire, Durham, New Hampshire

S. Mignard Institut Francais du Pétrole, Rueil-Malmaison, France

A. R. Mohamed Department of Chemical Engineering, University of New Hampshire, Durham, New Hampshire

Claude Moreau Laboratoire de Chimie Organique Physique et Cinétique Chimique Appliquées, Ecole Nationale Supérieure de Chimie, Montpellier, France

James Mudra Texaco Research and Development, Port Arthur, Texas

M. Neurock [†] Center for Catalytic Science and Technology, Department of Chemical Engineering, University of Delaware, Newark, Delaware

F. T. T. Ng Department of Chemical Engineering, University of Waterloo, Waterloo, Ontario, Canada

Tuan A. Nguyen Unocal Fred L. Hartley Research Center, Brea, California

Peter Nielsen-Hannerup Haldor Topsøe A/S, Nymøllevej, Lyngby, Denmark

G. Nongbri Texaco Research and Development, Port Arthur, Texas

Current affiliation: Merck Chemical Manufacturing Division, Rahway, New Jersey
[†]*Current affiliation*: Schuit Institute of Catalysis, Technical University of Eindhoven, Eindhoven, The Netherlands

Michael C. Oballa   Novacor Research and Technology Corporation, Calgary, Alberta, Canada

C. A. Paul   Texaco Research and Development, Port Arthur, Texas

R. J. Quann   Mobil Research and Development Corporation, Paulsboro, New Jersey

R. T. Rintjema   Department of Chemical Engineering, University of Waterloo, Waterloo, Ontario, Canada

A. I. Rodarte   Texaco Research and Development, Port Arthur, Texas

M. A. Salahuddin   Department of Chemical Engineering, University of New Hampshire, Durham, New Hampshire

D. E. Self   Texaco Research and Development, Port Arthur, Texas

Stuart S. Shih   Mobil Research and Development Corporation, Paulsboro, New Jersey

Milan Skripek   Unocal Fred L. Hartley Research Center, Brea, California

Kevin J. Smith   Department of Chemical Engineering, University of British Columbia, Vancouver, British Columbia, Canada

Peter Søgaard-Andersen   Haldor Topsøe A/S, Nymøllevej, Lyngby, Denmark

A. Stanislaus   Petroleum Technology Department, Kuwait Institute for Scientific Research, Safat, Kuwait

G. L. B. Thielemans   Shell International Petroleum Mij., The Hague, The Netherlands

J. P. van den Berg   Shell International Petroleum Mij., The Hague, The Netherlands

J. van Gestel   Laboratoire Catalyse et Spectrochimie, Université de Caen, France

H. M. J. H. van Hooijdonk   Shell International Petroleum Mij., The Hague, The Netherlands

Fritz Wenzel   VEBA OEL Technologie und Automatisierung GmbH, Gelsenkirchen, Germany

Chi Wong   Novacor Research and Technology Corporation, Calgary, Alberta, Canada

Sok M. Yui   Research Center, Syncrude Canada Ltd., Edmonton, Alberta, Canada

1 Catalyst Deactivation in Hydrotreating of Residua: A Review

C. H. Bartholomew

Department of Chemical Engineering
Brigham Young University
Provo, Utah 84602

INTRODUCTION

Hydrotreating, the catalytic conversion and removal of organic sulfur, nitrogen, oxygen and metals from petroleum crudes at high hydrogen pressures and accompanied by hydrogenation of unsaturates and cracking of petroleum feedstocks to lower molecular hydrocarbons plays an ever increasing key role in the refinery. Indeed, hydrotreating capacity has been growing steadily (at about 6% per year since 1976) and represents today nearly 50% of the total refining capacity (1). The increased application of hydrotreating can be ascribed to (i) the ever decreasing availability of light, sweet crudes and thus the increasing fraction of heavy, sour crudes that must be processed and (ii) the trend to increase upgrading of feedstocks for improvement of downstream processing such as catalytic reforming and catalytic cracking.

1

Hydrotreating of petroleum residua feedstocks involves three important reactions: hydrodesulfurization (HDS), hydrodenitrogenation (HDN), and hydrodemetallization (HDM) for removal of organically-bound sulfur, nitrogen, and metals respectively. Sulfided Mo, CoMo, and NiMo catalysts used in these reactions are prepared by impregnating catalyst extrudates with solutions of Co, Mo and Ni followed by drying, calcination at 400-500°C, and sulfiding with H_2S/H_2 at 350-400°C. The active sites for HDS and HDN are thought to be sulfur vacancies at the surface of a sulfide phase, e.g. Co_xMo_yS.

Hydrotreating involves a number of catalytic steps. For example, reaction steps in HDS include (see Fig. 1): (i) adsorptions of H_2 and the organic sulfide, (ii) hydrogenolysis of the carbon-sulfur bond, (iii) hydrogenation of unsaturates, (iv) hydrocracking, and (v) desorptions of hydrocarbons and H_2S. An important objective in hydrotreating is to maximize the rates of S, N, and metal removal, while minimizing the rates of hydrogenation and hydrocracking and therewith hydrogen consumption.

Sulfided resid hydrotreating catalysts are deactivated over a period of months by coke, metals and nitrogen compounds. The deactivation process involves a combination of uniform poisoning, pore mouth poisoning and pore blockage by (i) decomposition of organometallic compounds and (ii) buildup of soft coke and its transformation over a period of time to hard, crystalline coke. These problems are minimized by careful selection of guard beds, reactor design, and catalyst design; moreover, it is possible to regenerate coked catalysts with an oxygen burn.

Deactivation of hydrotreating catalysts has been fairly extensively studied (1-32). Several previous reviews of the literature (1-6) and an international symposium (7) have covered in some depth most aspects of this subject. Never-

Surface Structure of Molybdenum Sulfide

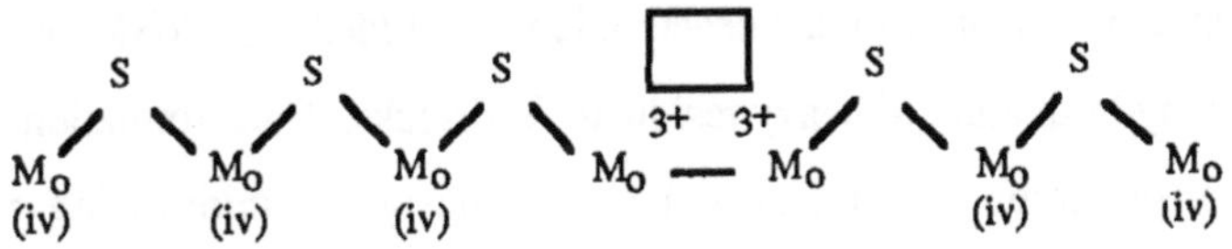

Reaction Steps in Hydrotreating

Adsorption of H_2 near vacancy (1)

Adsorption of RS on vacancy (2)

Hydrogenolysis of C-S bond (3)
$$RC\text{-}SH + H_2 \rightarrow RCH + H_2S$$

Hydrogenation of unsaturates (4)

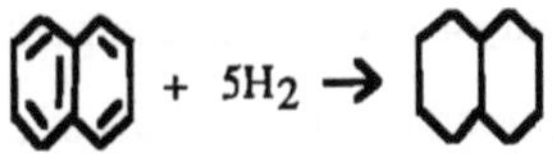

Hydrocracking (5)
$$RCH_2CH_2R' + H_2 \rightarrow RCH_3 + R'CH_3$$

Desorption of HC (6)

Desorption of H_2S to create vacancy (7)
$$H_2 + 2Mo^{4+} + S^{2-} \rightarrow \boxed{} + H_2S + 2Mo^{3+}$$

FIG. 1 *Reaction Steps in Hydrodesulfurization on a Sulfided Molybdenum Catalyst*

theless, an updated overview of the key aspects of residua hydrotreating deactivation, including coke formation chemistry, metals deposition chemistry, catalyst and reactor design, and the use of mathematical models to simulate the deactivation process may be timely.

This review focuses on the deactivation of sulfided Mo, CoMo, and NiMo catalysts in hydrotreating of heavy residuum feedstocks. Coke formation, metals deposition, the roles of catalyst and reactor design in minimizing catalyst decline, and the application of modeling to design and prediction of deactivation rates are discussed in this review in the sections which follow.

COKE FORMATION IN HYDROTREATING

The chemistry of coke formation and conversion in hydrotreating has been described by Beuther et al. (15); these same authors (15), Tamm et al. (21), and Simpson (32) have reported regarding the distribution of various types and overall concentrations of coke through the catalyst bed and in the pellets as a function of reaction time. Beuther et al (15) reported that three kinds of coke are formed in HDS (see Fig. 2): (i) Type I which includes strongly, but nevertheless reversibly adsorbed aromatic hydrocarbons from the feed such as benzene and naphthalene and polynuclear aromatics formed by polymerization, (ii) Type II which is formed by thermal uncoupling of asphaltic clusters and their strong adsorption on catalytic sites, and (iii) Type III involving the formation of a"hard coke" in the form of polynuclear aromatic "mesophase crystals."

Types I and II are formed early in the reaction cycle (see Fig. 3). Type I is thought to account for the initial rapid loss of activity requiring a relatively rapid increase in the reaction temperature during the first few weeks in order to maintain the desired conversion. Types I and II are slowly dehydrogenated, condensed and ultimately converted in part to Type III (end of run conditions in Fig. 3b) causing catastrophic loss of activity and ultimately plant shut-down. According to Thakur and Thomas (4), coke builds relatively rapidly to a maximum level within about 20% of the run cycle,   coinciding with the time at

a. Type I: Reversible adsorption and polymerization of aromatics:

(1) adsorption on vacancy or sulfide surface:

(2) polymerization

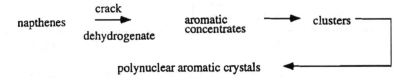

b. Type II: Thermal uncoupling of asphaltic clusters and adsorption on a surface site S:

c. Type III: Mesophase crystal formation:

$$napthenes \xrightarrow[\text{dehydrogenate}]{\text{crack}} \text{aromatic concentrates} \longrightarrow \text{clusters}$$

polynuclear aromatic crystals

FIG. 2 *Reactions in Hydrodesulfurization Causing Formation of Coke of Types I, II, and III (15)*

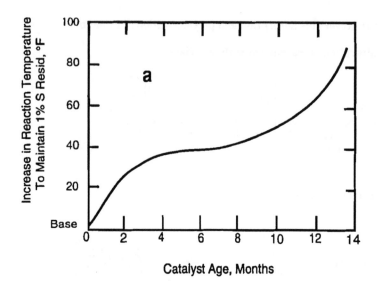

Catalyst Age, Months

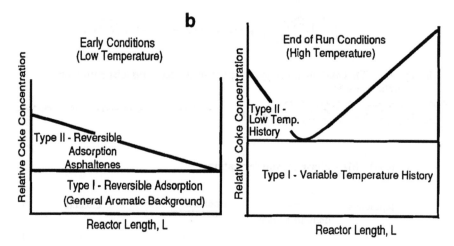

FIG. 3 *(a) A Typical Design Run for Hydrotreating of Kuwait Reduced Crude ;*
(b) Distribution of Coke Types during Early, Low-Temperature and Late Run,
High-Temperature Conditions (15)

which relatively constant activity is obtained and remains at that level throughout the remainder of the run (see Fig. 4). Since Type II coke is formed from asphaltenes, it follows that the extent of coke formation would depend on the asphaltene content of the petroleum feedstock. Thus, petroleum residua, which contain 6-11 wt.% asphaltenes, are feedstocks having a high tendency for coke formation.

Although it is reported that the concentration of coke formed in hydrotreating generally decreases through a fixed catalyst bed and from pellet edge to center consistent with a parallel reaction/deactivation network (4), data from Beuther and coworkers (15) in Fig. 5 show that coke formed from petroleum residua can be fairly uniformly deposited through the catalyst bed (in a fixed bed reactor), and inside the pellet, although it is evident from Fig. 5b that the deposit level is somewhat lower toward the outside of the pellet where the concentration of the deposited metals is highest. This can be attributed to the ability of the deposited V and Ni sulfides to gasify a portion of the deposited coke (21).

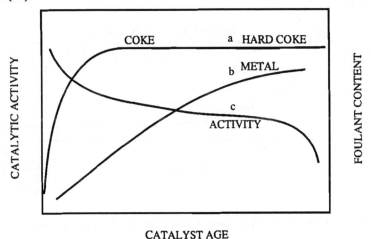

FIG. 4 *Changes in coke concentration, deposited metals concentration , and catalyst activity with time (4)*

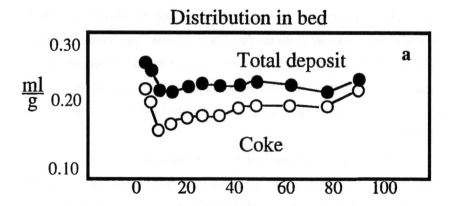

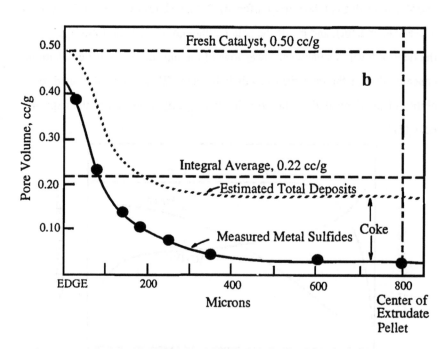

FIG. 5 *(a) Coke and total deposit distribution through catalyst bed; (b) Distribution of metals and coke in catalyst pellet (15)*

DEACTIVATION BY METALS

Petroleum residua typically contain 10-60 ppm Ni and 25-300 ppm V as organometallic compounds, e.g., porphyrins, coordinated with polynuclear aromatics, inside asphaltenes micelle clusters of 4-5 nm diameter (see Fig. 6). During high temperature reaction the organometallic compounds are decomposed on the catalyst surface and sulfided by gas phase H_2S. The transformation of the surface from a high activity Mo or CoMo sulfide to a Ni or V sulfide of low HDS activity effectively irreversibly poisons the surface. Thus, HDM is an irreversible adsorption rather than catalytic process. Because the decomposition process is rapid relative to the rate of pore diffusion, deposition occurs preferentially at entrances to reactors and catalyst pores. As the concentration of metals builds, pore mouth poisoning and blockage can occur. This has important implications for catalyst and reactor design that are discussed in a the next section.

Deactivation of hydrotreating catalysts by metals has been reviewed briefly by several authors (1-5) and in detail by Wei et al. (6, 33). Beuther et al.

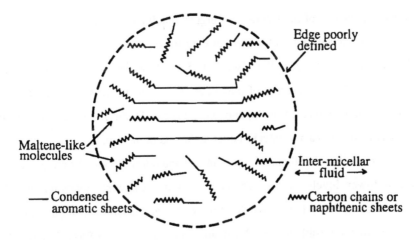

FIG. 6 *Schematic representation of an asphaltene cluster (3)*

(15), Tamm et al. (21) and Green and Broderick (22) have reported the results of definitive studies of metals deposition and its effects on catalyst activity.

According to Wei (33) the deposition of metals involves consecutive kinetics, i.e. hydrogenation of the metal-porphryin, e.g. nickel etioporphryin, followed by decomposition of the metal-chlorin:

$$NiEP \rightarrow NiEPH_2 \rightarrow Ni \text{ deposit} \qquad\qquad (1)$$

Pazos et al. (34), on the other hand, proposed a more complex combined series/parallel network. Overall rates of Ni and V metal deposition are reported to be first order (3).

Models of deactivation by metals have evolved in their complexity (33). For example, the model of metal sulfide deposit morphology has evolved from uniform layers to large crystallites and the model of support structure has evolved from uniform cylindrical pores to a detailed microstructure, e.g random spheres and needles. Some of the more recent, sophisticated models require the services of the fastest Cray (33).

Due to the low concentrations of metals in the fluid phase Ni, and V deposits generally accumulate slowly and gradually with time (see Fig. 4). However, because of the fast rates of organometallic compound decomposition , the deposits accumulate at the front of the catalyst bed (see Fig. 7a) and toward the outside of catalyst pellets (see Fig. 5b). That these deposits poison the catalyst, leading to an order of magnitude loss in activity, is evident from the data of Tamm et al (21, Fig. 7b) showing a gradual factor of 10 increase in HDS activity from the reactor inlet to outlet at the end of the run.

In addition to Ni and V, other metals including Pb and As are present at much lower levels in the feedstock and are likewise deposited on the hydrotreating catalyst, although As is more strongly adsorbed than Pb. Since Pb is somewhat reversibly adsorbed and since very low levels of Pb (> 2 ppb) in the gasoline boiling range hydrocarbons can result in poisoning of the downstream

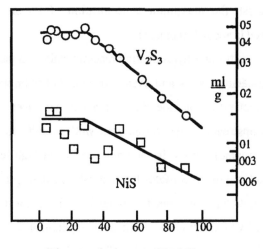

Distance from top of bed.%

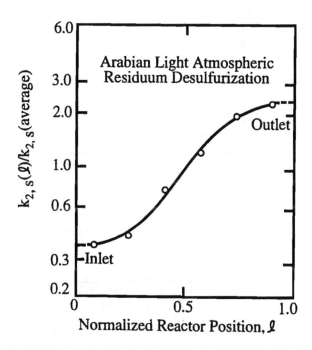

FIG. 7 *(a) Metals Deposit as a Function of Position in the Catalyst Bed (15) ;*
(b) Catalyst Activity at End of Run as a Function of Reactor Position (21)

reforming catalyst, the maximum concentration of Pb allowable for the hydrotreating feed is about 100 ppb (1).

Fe and Na salts enter hydrotreating reactors mainly as particulate matter which can plug catalyst pores and even beds (1). Other impurities that may contribute to deactivation include silica from antifoam agents introduced as various places upstream in the refinery as well as sulfates, and chlorides (inorganic and organic) present in the crude. Silica at high levels can cause blockage of catalyst pores. Sulfates lower crush strength and reduce activity following regeneration possibly due to formation of aluminum sulfates. Organic chlorides lead to HCl formation and downstream equipment corrosion.

Typical deposition patterns for Fe, Ni, and V determined by microprobe analysis following hydrotreating of an Arabian resid are shown in Fig. 8 accord-

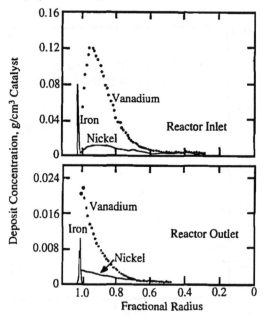

FIG. 8 *Typical Deposition Patterns for Fe, Ni, and V in a Hydrotreating Catalyst during Processing of Arabian Heavy Atmospheric Residuum at 700°F and a H₂ Partial Pressure of 1825 psia (1/16 inch Extrudate) (21)*

ing to Tamm et al. (21). It is evident that the deposition of Fe occurs on the outside of pellets as would be expected for a scale deposit containing relatively large particles. Deposits of Ni and V are found in a shell at a fraction radius of 0.6 to 1.0 at the reactor inlet and 0.8 to 1.0 at the reactor outlet. This behavior is consistent with a strong pore diffusional resistance for the demetallation process. The implications of the shell-like metals deposition profile for catalyst design are discussed in the next section.

CATALYST DESIGN: EFFECTS OF PORE STRUCTURE, SUPPORT AND CATALYST PROFILE ON ACTIVITY AND CATALYST LIFE

Hydrotreating catalysts are typically well-dispersed transition metal sulfides of either Mo or W promoted chemically with Co or Ni and supported on a γ-alumina carrier, sometimes mixed with silica. K, P, B, or rare earth oxide promoters may be added to lower the acidity of the catalyst and/or improve Mo sulfidability (35,36). Addition of P is known to increase the activity of Ni-Mo catalysts for HDN (36). Whether P increases Mo dispersion is controversial (35,36). The compositional range and physical properties of typical hydrotreating catalysts are listed in Table 1.

While Co and Ni sulfides by themselves have relatively low HDS or HDN activity relative to Mo sulfide, when combined in solid solution with Mo they serve to increase the specific activity of Mo by factors of 2-10. CoMo catalysts are generally more active for HDS, NiMo catalysts for HDN. The more expensive NiW catalysts are used principally for special applications in which high saturation and moderate cracking of low-sulfur feedstocks are desired (1).

A key aspect of catalyst design in hydrotreating is the optimization of pore structure. Support mesoporosity (pore diam. of 3-50 nm) is important for

TABLE 1 *Composition and Properties of Typical Hydrotreating Catalysts*

Composit./Properties[a]	Range	Typical Values
Active Phases (wt.%)		
MoO_3	13-20	15
CoO	2.5-3.5	3.0
NiO	2.5-3.5	3.0
Promoters (wt%)		
SiO_2	1-10	4
B, P	1-10	
Physical Properties		
Surface area (m^2/g)	150-500	180-300
Pore Vol. (cm^3/g)	0.25-0.8	0.5-0.6
Pore Diam (nm)		
mesopores	3-50	7-20
macropores	100-5,000	600-1,000
Extrudate diam.(mm)	0.8-4	3
Extrud. length/diam.	2-4	3
Bulk density (kg/m^3)	500-1,000	750
Ave. crush strength/ length (kg/mm)	1.0-2.5	1.9

[a]Catalyst is composed of active phases, promoters, and a γ-alumina carrier

providing and maintaining a well-dispersed sulfide layer; smaller pores provide a higher surface area for dispersing the active phase and higher surface area results in higher catalytic activity. On the other hand, pores must be large enough to admit large hydrocarbon molecules and metal-containing clusters but at the same time small enough to exclude asphaltenes and coke precursors, thereby minimizing coking. Pores must also be large enough to allow for deposition of carbon and metals while minimizing pore diffusional resistance and/or pore plugging. *These constraints require a careful optimization of the size distribution of mesopores in a resid hydrotreating catalyst with a compromise between high HDS activity and a low deactivation rate.*

Four different mesopore (sometimes loosely referred to as micropore) designs for hydrotreating catalysts are illustrated in Fig. 9: (1) a sharp pore diameter distribution around 7-10 nm, (2) a sharp pore diameter range from 10-20 nm, (3) a fairly broad pore diameter distribution from about 10-50 nm, and (4) a bimodal pore size distribution with meso and macropores. The first design with small pores is typical of distillate HDS catalysts, while recently developed resid desulfurization (RDS) catalysts have mesopores in the 10-20 nm range as illustrated by the design of Curve 2. HDM catalysts have large mesopores as in Curve 3.

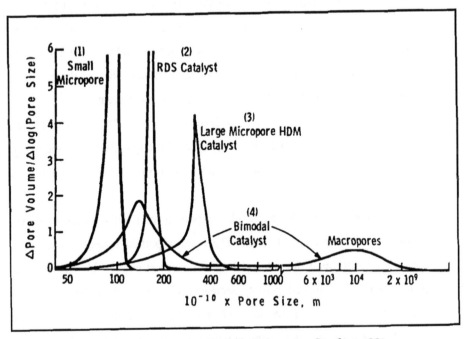

FIG. 9 *Pore Size Distributions of Typical Hydrotreating Catalysts (22)*

Hydrotreating catalysts including RDS catalysts are often designed with a bimodal pore size distribution with maxima in the meso and macropore regions such as in Curve 4 (Fig. 9). Support macroporosity is desirable to minimize pore diffusional resistance--i.e. maximize the access of reactants to the active surface and of asphaltenes and other large molecules to the interior of the catalyst pellet. Total porosity, however, is typically limited to 0.5-0.6 or less to maintain pellet strength, since strength decreases with increasing porosity (37).

Although a number of previous studies show that hydrotreating catalysts with bimodal pore size distributions result in longer catalyst life than those with unimodal distributions, Absi-Halabi et al. [38] recently reported that the activity and resistance to deactivation of a resid hydrotreating catalyst with a broad unimodal distribution (pore diameters between 5 and 50 nm with 70-80% in the desired mesopore range) exceed those of selected bimodal catalysts.

Another aspect of catalyst design that influences catalyst activity and life is the profile of active material in the catalyst pellet, which can be varied by controlling the preparation procedure. Since during HDM Ni and V deposits which deactivate the catalyst are concentrated in an outer shell of the pellet (or extrudate), there is a clear advantage to preparing hydrotreating catalysts for treatment of metal-containing feeds with the active phase "center loaded," i.e., concentrated at an "inner shell" or "yoke" approximately halfway into the pellet. The advantage of such a catalyst distribution is also predicted by modeling of the HDM process (33).

Catalyst size and shape also play important roles in the design of active, stable hydrotreating catalysts (22). There are important tradeoffs in activity, pressure drop through the catalyst bed and crush strength with variations in pellet size (22,35). Indeed, activity decreases with increasing pellet size due to higher pore diffusional resistance, while pressure drop decreases and crush strength increases with increasing pellet size. Thus, to maximize activity and crush strength while minimizing pressure drop, the pellet size must be optimized. Table 1 shows that extrudate diameters are typically small (0.8-4.0 mm) in order to minimize pore diffusional resistance. Shaped extrudates, e.g. stars, trilobes and quadralobes, are commonly used in hydrotreating applications, since they provide the advantages of high geometrical surface area, low pressure drop, high crush strength and high contaminant metals tolerance (35,39).

From the foregoing discussion it should be clear that the physical properties of hydrotreating catalysts may change very significantly over a run cycle. Inoguchi et al. (40) have published data on the effects of catalyst deactivation on surface area, pore volume and mean pore diameter (see Table 2). From these data it is evident that surface areas and pore volumes decrease more than 50% and mean pore diameters decrease about 15-20% over a run cycle. Only 10-60% of the lost surface area and pore volume are restored during regeneration; less is recovered in the upper part of (or entrance to) the catalyst bed probably as a result of metals poisoning.

Data based on N_2 adsorption measurements from Beuther and coworkers (28) in Figure 10 provide a breakdown of surface area with pore radius for a resid hydrotreating catalyst, while Fig. 11 illustrates changes in pore volume with pore radius and effects of coke deposition on the accessible pore volume. It is evident from Fig. 10 that most of the surface area in the fresh catalyst is distributed in the smaller mesopores with radii of 2 to 5 nm. Similarly, most of the pore volume of the fresh catalyst measured by N_2 adsorption occurs in pores with radii of 2 to 5 nm. Comparison of Figures 11b and 11c indicates that about 50% of the mesopore volume in the deactivated catalyst is taken up by carbon deposits. Moreover, there is an obvious shift in the mean pore diameter from about 35 nm to about 25 nm as a result of coking. This shift has been attributed to adsorbed asphaltene micelles in the larger pores (15).

TABLE 2 *Physical properties of fresh, used and regenerated hydrotreating catalysts; Catalyst age = approximately 3000 h, feed = Kuwait Resid (40)*

Catalyst	Condition	Surface area (m^2g^{-1})	Pore Volume $(ml\ g^{-1})$	Mean pore diameter (nm)
KC 28	Fresh	239	0.458	76
	Spent	106[a]	0.172	65
		102[b]	0.187	74
	Regener.	119[a]	0.233	78
		170[b]	0.343	81
KS 83	Fresh	128	0.367	115
	Used	55[a]	0.144	106
		50[b]	0.140	113
	Regener.	69[a]	0.197	115
		81[b]	0.259	127

[a]Upper part of catalyst bed.
[b]Lower part of catalyst bed.

EFFECTS OF PROCESS CONDITIONS, FEED PRETREATMENT, AND REACTOR DESIGN ON CATALYST DEACTIVATION

While feedstock composition and catalyst design play key roles in the deactivation process, process conditions, feed pretreatment and reactor design play equally crucial roles in determining catalyst life and processing efficiency (4).

The most important process variables in hydrotreating catalyst deactivation are temperature, H_2 pressure, and space velocity. Coke formation and transformation and metals deposition rates increase with increasing temper-

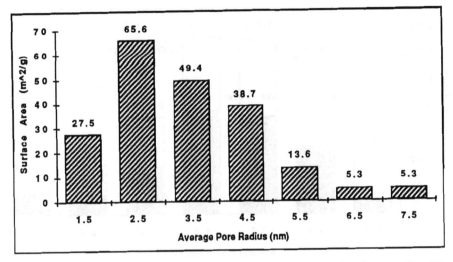

FIG. 10 *Distribution of Surface Area of a Hydrotreating Catalyst by Pore Size (5)*

ature. At elevated reaction temperatures (>400°C) coke and metal deposits concentrate near the front of the active reaction zone in the catalyst bed and near pellet edges causing more rapid plugging of catalyst pores. The formation of hard coke is also accelerated under these severe conditions; indeed this accounts for the acceleration in deactivation rate and likewise temperature required for maintaining constant conversion typically observed at end of run conditions. In other words, it is the acceleration of coke hardening and pore mouth plugging rates with increasing temperature that forces termination of the run.

Increasing H_2 partial pressure decreases the rates of coke formation and hardening and extends catalyst life, if the activity loss is principally due to coke deposition (4). On the other hand, metals deposition rates, particularly at the entrance to the reactor and at the outer edges of catalyst particles, increase with increasing H_2 partial pressure; accordingly, if deactivation by metals dominates, lower H_2 partial pressures are preferred (4). At otherwise constant reaction conditions in a reactor of a given type, lower rates of deactivation are observed at lower space velocities (4).

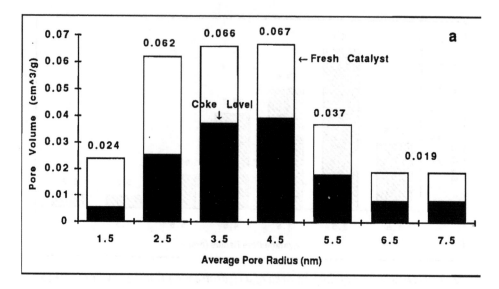

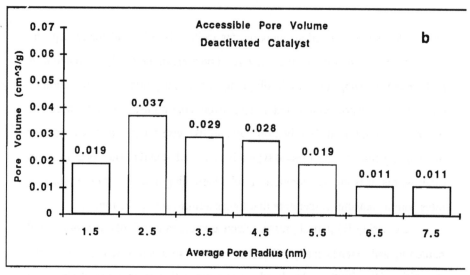

FIG. 11   a. Distribution of Pore Volume of a Hydrotreating Catalyst and Coke
Levels in the Deactivated Catalyst with Pore Size; b. Accessible Pore
Volume in the Deactivated Catalyst (15)

According to Thakur and Thomas (4) "any viable process for treating heavy [petroleum or coal-derived] feeds should have provisions for feed pretreatment." These may involve (i) extraction methods to remove asphaltenes and metals , (ii) other types of preprocessing (e.g. visbreaking, coking, and/or hydrocracking), or (iii) catalytic demetallation using an inexpensive catalyst. Fig. 12 illustrates the concept of a staged reactor involving three beds of catalysts of increasing pore size, activity and cost. Metals are removed in large part in the first bed containing a large-pore, throw-away Mo/alumina catalyst; the remaining metals are removed and HDS is initiated in the medium-pore CoMo/alumina catalyst of moderate activity, while the remaining sulfur is converted in the small-pore, high-activity CoMo/alumina catalyst. Fig. 13a shows how a two-stage reactor system enables higher sulfur conversion with increasing catalyst age relative to a single-stage process. The application of a guard bed to substantially lower Fe and As concentrations in a hydrotreating catalyst is illustrated in Fig. 13b.

Two kinds of hydrotreating reactors, fixed and ebullated (or expanded slurry), are used in the treatment of heavy hydrocarbon feedstocks (4). The fixed bed, which has been used more extensively, has the advantages of (i) better temperature control, (ii) wider operating flow range, and (iii) little or no catalyst attrition and/or agglomeration but the disadvantages of (i) non-isothermal operation with the requirement of a quench system to limit the temperature increase across the bed, (ii) gradients of coke, metals, and poisons, and (iii) high pore diffusional resistance and mass transport limitations. The merits of the ebullated or slurry bed include: (i) isothermal operation with no quenching and therewith less coking, (ii) elimination of mass transport limitations, (iii) no blockage of flow as in a fixed bed, and (iv) more uniform deposition of coke and poisons across the bed and catalyst particles, thereby minimizing plugging of the catalyst, although metal gradients in catalyst pellets are still observed (4). On

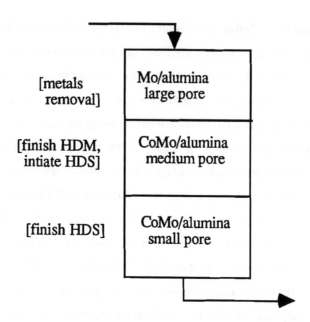

FIG. 12 *Staged Reactor System with Decreasing Pore Size Strategy for HDM/HDS of Residua*

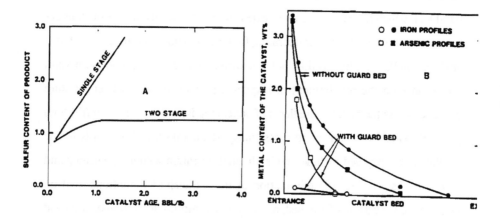

FIG. 13 *a. Sulfur content of HDS product for Two-Stage and One-Stage Processes; b. Fe and As Metal Contents with and without Guard Bed (4)*

the downside the ebullating bed (i) must be monitored closely to prevent temperature runaway, (ii) suffers from catalyst attrition and agglomeration, and (iii) requires sophisticated equipment for separating the catalyst fines from the product.

MODELING OF HDS/HDM IN RESIDUA HYDROTREATING

Modeling of deactivation processes is an important activity that can provide powerful predictive tools for simulation of processes over a range of scales. A comprehensive, robust model can provide (i) the capability to simulate complex processes at a large scale at a fraction of the cost of operating demonstration units, (ii) useful insights into the design of catalysts and reactors, and (iii) the prediction of catalyst life. Such models are presently being used in the petroleum industry at the refinery to predict the useful life of hydrotreating catalysts. With the advent of relatively inexpensive, sophisticated computers the use of such models is increasing.

Several modeling studies of hydrotreating processes have been reported (13, 16-18, 24, 25, 27, 30, 41). While these models vary in their approach and sophistication, most share common elements: (i) a simplified reactor design equation, (ii) simplified rate expressions for the main reactions, (iii) deactivation rate equations for the deactivation processes, (iii) equations to account for pore diffusional resistance in the main and deactivating reactions, and (iv) a simplified method of representing the hydrocarbon feed (e.g. a single representative molecule, lumped parameters for different molecular types, or molecular weight distributions).

The work of Nitta et al. (13) serves to illustrate the development of a relatively simple but nevertheless, effective model for HDS of a heavy oil. The basic features of this model include: (i) second order kinetics with high pore diffusional resistances for HDS and HDM of V (first order in H_2 for both reactions), (ii) lumping of Ni deposition with V, (iii) reversible coke deposition with second order kinetics (first order each in H_2 and coke) for the reverse step, (iv) a hindered diffusional model accordingly to Spry and Sawyer (42) which incorporates a decrease in diffusivity proportional to the volume of deposited V and coke, and (v) an intraparticle model for the catalyst particles which is incorporated into the design equations for a plug flow reactor. The data input includes initial catalyst activity for HDS. Experimental measurements of temperature, sulfur product concentration and V in the product during HDS of an Iranian heavy vacuum resid at the bench scale are compared with model simulations of the same parameters in Fig. 14. Very good agreement of experiment and model is evident for the set of conditions investigated.

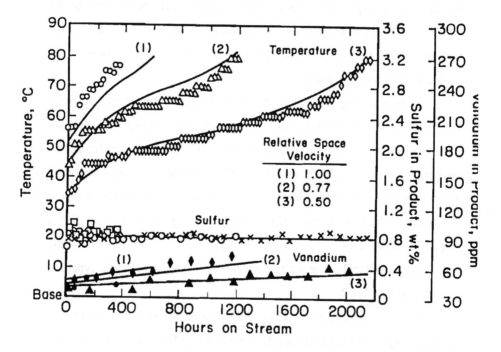

FIG. 14 *Model Simulation of HDS of an Iranian Heavy Vacuum Resid on a Bench-Scale Fixed Bed Reactor (13)*

While its capability for simulating resid HDS is apparently very good , the model of Nitta et al. could be improved by incorporating (i) a feedstock size distribution and reactivity correlation, (ii) a rate equation for coke transformation (hardening), (iii) a more generalized model for catalyst pore structure, and (iv) a generalized correlation for catalyst activity with Co and Mo dispersion.

A more sophisticated approach to modeling of residua HDM has been discussed by Sughrue et al. (41). Their approach incorporates:

(i) a vanadium molecular size distribution measured by size exclusion chromatography as an input for reactor modeling;

(ii) a rate of deposition of metal sulfide mass (m) per unit length as

$$dm/dt = \alpha \, r_p \, k_s \, M \, C \qquad (2)$$

where α is the number of metal sulfide molecules per molecule of organometallic reactant, r_p is the pore radius at any given time, k_s is the surface rate constant defined below, M is the molecular weight of the metal sulfide, and C is the concentration of V; and

(iii) a surface rate constant modified for deactivation by metals as follows:

$$k_s = k_{so}(1-m/m_s) + k_{sc}(m/m_s) \qquad (3)$$

where k_{so} is the rate constant for a fresh catalyst surface, k_{sc} is the rate constant for a V-covered surface, and m_s is the mass corresponding to complete active site poisoning. When m is greater than m_s, k_s is equal to k_{sc}.

Finally, after accounting for changes in the molecular size distribution due to thermal effects (i.e., thermal cracking), the agreement of experimental and predicted product distributions are excellent (see Fig. 15).

The effects on model predictions of temperature versus time of using a feed molecular size distribution versus an average molecular size were also considered by Sughrue et al. (41).   Their temperature-time data for three differ-

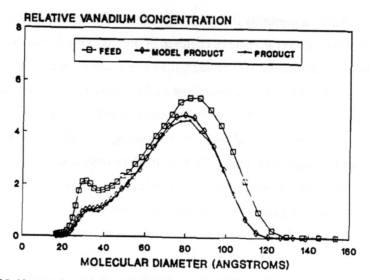

FIG. 15 *Vanadium Molecule Size Distributions of Feed and Product (Measured and Predicted) for Hondo Asphaltenes (41)*

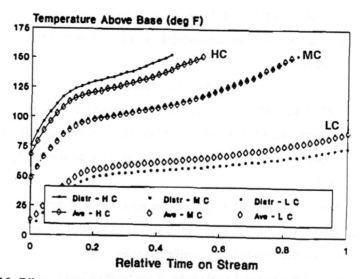

FIG. 16 *Effects on HDM temperature-time model predictions of using average molecular size versus a size distribution as input at high, moderate, and low conversion levels (41)*

ent conversion levels in Figure 16 show that use of an average molecular size at high conversion results in prediction of a lower start-of-run temperature and a longer catalyst life, while the reverse is true at low conversion. At medium conversion level, the two methods agree in their prediction.

CONCLUSIONS

1. Deactivation of hydrotreating catalyst occurs mainly by coking and metals deposition. Saturation coke levels are reached early in a typical run cycle, although a slow transformation from soft to hard coke occurs gradually along with metals deposition over the course of the run cycle. Coke which comprises the large part of deposits on hydrotreating catalysts may concentrate in fixed beds at the reactor entrance and pellet edges, although it is probably more often uniformly spread through pellet and catalyst bed, Metals, on the other hand, are concentrated at the exterior of pellets and at the entrance to the catalyst bed.

2. Deactivation rate is a function of feedstock and catalyst properties, reactor type, reaction conditions, and feed pretreatment.

3. Key feedstock properties that affect deactivation rate include (i) asphaltene content, (ii) V and Ni metals content, and (iii) significant concentrations of poisons, e.g., nitrogen compounds, As, and Pb and foulants, e.g. Fe, sulfates, and silica.

4. Catalytic properties affecting deactivation rate include acidity, catalyst distribution on pellets, pore size distribution, and surface area. In catalyst design there are important tradeoffs in the choice of these catalysts properties, e.g., increasing pore size decreases deactivation rate but at the expense of lower surface area and initial activity. Thus, pore size distribution is optimized for different functions, i.e. large pores for metals

removal and smaller pores for hydrodesulfurization. Both meso and macroporosity are important in facilitating diffusion of large molecules in and out of the catalyst during reaction.

5. Important process variables in hydrotreating catalyst deactivation are temperature, H_2 pressure, and space velocity. Coke formation-/transformation and metals deposition rates increase with increasing temperature. At elevated reaction temperatures coke and metal deposits concentrate near the front of the active reaction zone in the catalyst bed and near pellet edges causing more rapid plugging of catalyst pores. The formation of hard coke is also accelerated under more severe conditions which ultimately forces termination of the run. Increasing H_2 partial pressure decreases the rates of coke formation/hardening but increases metals deposition rates. Lower rates of deactivation occur at lower space velocities.

6. Feed pretreatment by either extraction of asphaltenes or catalytic metals removal is essential for effective hydrotreating of residua. Staged reactors with beds of progressively smaller pore size and higher activity facilitate efficient, economical removal of sulfur and metals and prolong catalyst life.

7. Both fixed and ebullated, slurry beds are used for hydrotreating of residua, although fixed beds are more common. The fixed bed has better temperature control, a wider range of operating conditions with little or no attrition but probably higher deactivation rates and lower catalyst life because of temperature, coke and metal gradients and mass transport limitations. Merits of the ebullated bed include isothermal operation and elimination of mass transport limitations and coke/metal gradients with attendant lower rates of deactivation; however, potential temperature instability, catalyst attrition and agglomeration, and the difficulty in separating catalyst fines from the product are disadvantages.

8. Modeling of deactivation requires an understanding of the reaction and deactivation kinetics; it can provide (i) useful insights into the design of catalyst pellets and reactors, (ii) simulation of processes on a large scale at a fraction of the cost of operating demonstration units, and (iii) quantitative prediction of catalyst life in a large scale process based on catalyst and feedstock properties. Hydrotreating models are already used in refineries to predict catalyst life, and their use is increasing.

ACKNOWLEDGMENTS

The author gratefully acknowledges financial support by Brigham Young University and useful discussions with Drs. Dennis H. Broderick and H. F. Harnsberger of Chevron Research and Technology Co., Dr. Anthony J. Perrotta of Gulf Research, Dr. Rydan L. Richardson of Unocal Research, Dr. Edward L. Sughrue of Phillips Petroleum Co., and Dr. Kenneth L. Riley of Exxon Research and Development Laboratory.

REFERENCES

1. McCulloch, D. C.; "Catalytic Hydrotreating in Petroleum Refining," in Applied Industrial Catalysis, ed. B. E. Leach, Academic Press, 1, 69, 1983.

2. Ward, J. W.; Qader, S. A.; Hydrocracking and Hydrotreating, ACS Symp. Series, 20, ACS, Washington, D. C., 1975.

3. Gates, B. C.; Katzer, J. R.; Schuit, G. C. A.; Chemistry of Catalytic Processes, Chapter 5, McGraw-Hill, 1979.

4. Thakur, D. S.; Thomas, M. G.; "Catalyst Deactivation in Heavy Petroleum and Synthetic Crude Processing: A Review," Appl. Catal., 15, 197-225, 1985.

5. Butt, J. B.; Petersen, E. E.; Activation, Deactivation, and Poisoning of Catalysts, Academic Press, New York, 1988.

6.   Quann, R. J.; Ware, R. A.; Hung, C.; Wei, J.; in Advances in Chemical
      Engineering, ed. James Wei, Academic Press, 14, 1988.

7.   Bartholomew, C. H.; Butt, J. B.; Catalyst Deactivation, Studies in Surface
      Science and Catalysis No. 68, Proceedings of the 5th International
      Symposium, Elsevier, 1991.

8.   Wheeler, A.; Robell, A. J.; J. Catal., 13, 299-305, 1969.

9.   Richardson, R. L.; Alley, S. K.; "Consideration of Catalyst Pore Structure
      and Asphaltenic Sulfur in the Desulfurization of Resids, in Hydrocracking
      and Hydrotreating, ACS Symp. Series, 20, ACS, Washington, D. C.,
      1975.

10   Prahser, B. D.; Ma, Y. H.; AIChE J., 23(3), 303-311, 1977.

11.   Parking, F. S.; Paraskos, J.; Frayer, J. A.; AIChE Symp. Series, 71(148),
      241-246.

12.   Rojagopalan, K.; Luss, D.; Ind. Eng. Chem. Process Des. Dev. 18(3),
      459-465, 1979.

13.   Nitta, H.; Takatsuka, T.; Kodama, J. S.; Yokoyama, T.; "Modeling of the
      HDS Process," AIChE. Natl. Mt., Houston, 1979.

14.   Riley, K. L.; Silbernagel, B.; "Effects of Metals and Asphaltene Content
      on HDS Activity and Metals Removal," Stud. Surf. Sci., Catal. 6, 313,
      1980.

15.   Beuther, H.; Larson, O. A.; Perrotta, A. J.; in Catalyst Deactivation, eds.
      B. Delmon and G. Froment, Elsevier, 271, 1980.

16.   Kodama, S.; Nitta, H.; Takatsuka, T.; Yokohama, T.; J. Japan Petr. Inst..
      23(5), 1980.

17.   Higashi, H.; "How to Estimate the Life of Hydrodesulfurization Catalysts
      for Heavy Oils," Petrotech, 3, 1980.

18.   Sie, S. T.; "Catalyst Deactivation By Poisoning and Pore Plugging in
      Petroleum Processing," in Catalyst Deactivation, eds. B. Delmon, and G.
      F. Froment, Elsevier Sci., Amsterdam, 1980.

19.   Langhout, W. C. V. Z.; Ouwerkerk, C.; Prank, K. M. A.; "Development
      and Experience with Shell Residue Hydroprocess," Prep. AIChE, 88th

AIChE Mt., June, 1980.

20. El-Kady, F. Y. A.; Mann, R.; J. Catal, 69, 147-157, 1981.

21. Tamm, P. W.; Harnsberger, H. F.; Bridge, A. G.; Ind. Eng. Chem.,
     Process Des. Dev., 20(2), 262-273, 1981.

22. Green, D.C.; Broderick, D.H.; Chem. Eng. Prog., 77, 33, 1981.

23. Shimura, M.; Shiroto, Y.; Takeuchi, C.; "Effect of Catalyst Pore Structure
     on Hydrotreating of Heavy Oil," Abst. ACS Div., Coll., 183 ACS Mtg.,
     March, 1982.

24. Hannerup, P. N.; Jacobsen, A. C.; " A Model for Deactivation of Residue
     Hydrodesulfurization Catalysis," Petr. Chem. Div., 185th National ACS
     Met., Seattle, Wash., March, 1983.

25. Nalitham, R. V.; Tarrer, A. R.; Guln, J. A.; Curtis, C. W.; "Application of
     a Catalyst Deactivation Model for Hydrotreating Solvent Refined Coal
     Feedstocks," Ind. Eng. Chem. Process Des. Dev. 22, 645-653, 1983.

26. Newson, E. J.; Ind. Eng. Chem. Process, Des., Dev. 14, 27, 1985.

27. Arteaga, A.; Fierro, J. L. G.; Delannay, F.; Delmon, B.; "Simulated
     Deactivation and Regeneration of an Industrial CoMo/γ-Al$_2$O$_3$
     Hydrodesulphurization Catalyst," Appl. Catal., 26, 227-249, 1986.

28. Chang, H. J.; Crynes, B. L.; "Effect of Catalyst Pore and Pellet Sizes on
     Deactivation in SCR Oil Hydrotreatment," AIChE J. 32, 224, 1986.

29. Yamamoto, Y.; Kumata, F.; Massoth, F. E.; "Hydrotreating Catalyst
     Deactivation by Coke from SCR-II Oil," Fuel Proc. Tech. 19, 253-263,
     1988.

30. Melkote, R. R.; Jensen, K. F.; "Models for Catalytic Pore Plugging:
     Application to Hydrodemetallation," Chem. Eng. Sci. 44, 649-663, 1989.

31. Myers, T. E.; Meyers, B. L.; Lee, F. S.; Fleisch, T. H.; Zajac, G. W.; "
     Resid Catalyst Deactivation in Expanded-Bed Service," Paper No. 5e,
     April 1989 Meeting.

32. Simpson,H. D.; "Aspects of Coke Deactivation in Hydroprocessing
     Catalysts," 1990 AIChE Annual Meeting.

33. Wei, J.; in C. H. Bartholomew and J. B. Butt, Catalyst Deactivation,

1991, Studies in Surface Science and Catalysis No. 68, Proceedings of the 5th International Symposium, Elsevier, 1991.

34.   Pazos, J. M.; Gonzalez, J. C.; Salazar-Guillen, A. J.; Ind. Eng. Chem. Proc. Des. Dev., 22, 653, 1983.

35.   Richardson, J. T.; Principles of Catalyst Development, Plenum Press, 1989.

36.   Prins, R.; "Supported Metal Sulfides," in Characterization of Catalytic Materials, ed. I. E. Wachs, Butterworth-Heinemann, 1992.

37.   Sleight, A. W.; Chowdry, U.; "Catalyst Design and Selection," in Applied Industrial Catalysis, ed. B. E. Leach, 2, 1.

38.   Absi-Halabi, M.; Stanislaus, A.; Al-Zaid, H.; Am. Inst Chem. Engr., Spring National Meeting, Houston, Texas, March 23-April 1, 1993.

39.   Richardson, R. L.; Riddick, F. C.; Ishikawa, M.; Oil Gas J., 77, 22, 1979.

40.   Inoguchi, M. et. al.; Bull. Japan Petrol. Inst., 14, 153, 1972.

41.   Sughrue, E. L.; Adarme, R.; Johnson, M. M.; Lord, C. J.; Phillips, M. D.; in Catalyst Deactivation 1991, eds. C. H. Bartholomew and J. B. Butt, Elsevier, 281, 1991.

42.   Spry, Jr., J. C.; Sawyer, W. H.; 68th Annual AIChE Meeting, Los Angeles, Paper 30C, 1975, .

2   Catalyst Deactivation in Residue Hydrocracking

Michael C. Oballa, Chi Wong and Andrzej Krzywicki

Novacor Research & Technology Corporation,

2928-16th Street, N.E., Calgary, Alberta, CANADA T2E 7K7

ABSTRACT

The existence of a computer-controlled bench scale hydrocracking unit at our site has made cheaper the non-stop running of experiments for long periods of time.   It was, therefore possible to show, at minimal costs, when three hydrocracking catalysts in service reach their maximum lifetime.

Different parameters which are helpful for catalyst life and activity predictions were calculated, e.g., relative catalyst age and the effectiveness factor.   Experimental results compared well with model, giving us the minimum and maximum catalyst lifetime, as well as the deactivation profile with regard to sulfur and metals removal.   Reaction rate constants for demetallation and desulfurization were also determined.   Six commercial catalysts were evaluated at short term runs and the three most active were used for long term runs.   Out of three catalysts tested for deactivation at long term runs, it was possible to choose one whose useful life was higher than the others.

All runs were carried out in a Robinson-Mahoney continuous flow stirred tank reactor, using 50/50 volumetric mixture of Cold Lake/Lloydminster atmospheric residue and $NiMo/Al_2O_3$ catalyst.

INTRODUCTION

Over the past few years, there has been a dwindling of reserves of light and medium crudes all over the world with North America being at the forefront. As a result, the world is going to rely more and more on heavy crudes and residual oils to meet the demand for valuable refined products and petrochemical feedstocks. It is in reply to these needs that Husky Oil and three other partners built the Bi-Provincial Upgrader (BPU) in Lloydminster to produce 7300 m^3/cd of synthetic crude. The core of the BPU is the Primary Upgrading Plant consisting of the H-OIL Unit, a Delayed Coker and a Gas Recovery Unit. The H-OIL Unit is designed to process 5088 m^3/sd of a 50/50 volumetric blend of Lloydminster and Cold Lake atmospheric residue (399°C+) in an ebullated bed hydrocracker [1].

The heavy crudes and residues however present more processing difficulties than light and medium crudes. This is a result of their higher content of metals (V, Ni, Fe), heteroatoms (S, N, O), and lower hydrogen to carbon ratio as a general rule. The metals and sulfur deposit on the surface of the catalyst thereby causing a deterioration of catalyst performance, i.e., deactivation. Catalyst deactivation is of major importance in the design or operation of reactors for the hydroprocessing of residue or heavy oils.

There have been different studies on catalyst deactivation in residue and heavy oil hydroprocessing [2-10] but the ones that have gained general practical industrial acceptance are based on the concept of pore plugging. The pore plugging concept was first proposed by Hiemenz [11] in 1963. His concept was based on gas permeability measurements on fresh and used catalysts. Beuther and Schmid [12] showed that the coke content of the catalyst increased very rapidly to an equilibrium value during the early stages of a run (40 hours in their experiment). Thereafter, the coke content of the catalyst seemed to remain constant with run time. Newson [13, 14] modelled deactivation by a pore-plugging mechanism based on a decrease of the effective diffusivity of the catalyst particles by intra-particle deposits in combination with coking effects. Studies by Shell Research Laboratories in Amsterdam [15, 16, 17] confirmed the coke deposition and pore plugging concepts and the findings of Hiemenz, Beuther and Schmid, and of Newson. The studies showed that after the initial rapid catalyst coking stage, the deposition of trace metals on the catalyst surface generally occurs slowly over the "active" life of the catalyst, followed by a rapid decline of catalyst activity. This last period is attributed to pore blockage and, in literary terms, is the death-knell of the catalyst.

Our primary objective was to choose the right and best commercial catalyst for hydrocracking the feedstock for the Bi-Provincial Upgrader. Not only the initial activity is to be considered, but also primarily, the longer term activity of the catalyst (the deactivation profile) should be known. The design feedstock and product properties are given in Table 1.

Table 1: *H-Oil Unit Feed and Product Properties*

		FEED	PRODUCT
Gravity	[API]	7.4	24.2
Hydrogen	[wt%]	10.38	11.82
Carbon	[wt%]	83.4	86.7
Sulfur	[wt%]	4.85	0.77
Nitrogen	[wt%]	5600	3270
Vanadium	[wppm]	206	41
Nickel	[wppm]	89	24

The choice had to be made based on actual runs starting with six commercial hydrocracking catalysts. Three of these catalysts were then used for long term runs. Differences in the activity of the catalysts were observed. This paper presents and discusses the results obtained from the runs as well as those from calculations based on run data.

EXPERIMENTAL

Feedstock: The feed is a 50/50 volumetric blend of Cold Lake and Lloydminster atmospheric residue (399°C+).

Catalysts: The catalysts are all of the Ni-Mo type on Alumina base and were obtained from commercial catalyst vendors. To protect the identity of the vendors, the letters A to F are used to identify the catalysts instead of the vendors' names. Each catalyst was dried *ex-situ* and sulfided *in-situ* with DMDS in gas oil before a run.

Analysis: Densities were determined at 15.5°C on a Paar DMA 48 Densitometer while dynamic viscosities were determined at 40°C on a Brookfield DV II viscometer. Sulfur in wt% was determined on a Leco SC-132 sulfur analyzer. Trace nitrogen was determined by chemiluminescence method on an Antek analyzer, while trace metals in the oils were determined by X-Ray fluorescence using a Horiba MESA 710 analyzer. Both simulated distillation and vacuum distillation were performed according to ASTM D-2887 and ASTM D-1160, respectively. Surface area of catalysts, pore volume and pore size distribution were measured using an Autoscan 60 Mercury Porosimeter. Coke on catalyst after a run was determined by coke burn-off in a muffle furnace at 550°C. Metals content of both fresh and spent catalysts were determined by Proton Induced X-Ray Emission (PIXE) method.

The Reaction Unit: The results reported here were obtained from a continuous flow reaction unit, designed and built in-house, and shown schematically in Figure 1. The system comprises of four major modules: gas and liquid feed section, reactor section, product separation and collection section, and the process control section.

In the gas feed section, the hydrogen is supplied from gas cylinders to a compressor. The compressed gas is stored in a surge drum. The flow rate of the gas from the surge drum as well as the pressure are controlled. The residuum is charged to a feed tank in a batch mode, heated to 130°C to reduce the viscosity, and transferred by a metering pump to the reactor inlet.

At the reactor inlet, the residuum is mixed with the high pressure hydrogen before entering the bottom of the reactor. The hydrocracking runs are carried out in a one liter 316 stainless steel Robinson-Mahoney type continuous stirred tank reactor. The required quantity of catalyst is fixed in a basket in the reactor and appropriate liquid flow rates are applied to give a catalyst space velocity and liquid hourly space velocity close to those of the

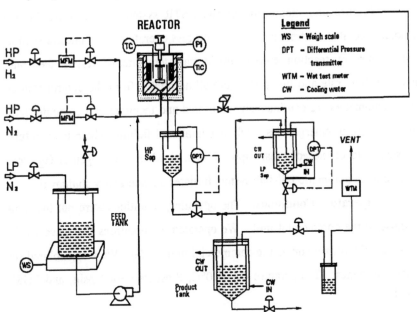

Figure 1: *Simplified Scheme of Reaction Apparatus*

commercial unit. Isothermal operation is maintained. The way the catalyst wire gauze basket is fixed in the reactor ensures that catalyst properties are the same throughout the bed at any time during a run and therefore the catalyst deactivates uniformly. It is very much unlike a fixed bed reactor which exhibits a deactivation front which moves smoothly through the bed depth.

The mixture of gaseous and liquid reaction products are transferred from the reactor through an overflow tube to a high pressure separator. The separator volume is one liter and liquid level is maintained through a pressure letdown valve connected to a level controller and transmitter. The gases from the high pressure separator go to a low pressure separator through a back pressure regulator. The liquid streams from both the high and low pressure separators are recombined to go into a product tank. The off-gas from the low pressure separator also goes into the product tank. This tank is kept at 15°C and its off-gases are scrubbed before venting through a wet test meter.

Control and monitoring of the CSTR system is performrd from a remotely operated control station. The process operational parameters such as system pressure, reaction temperature, furnace temperature, residue feed rate, hydrogen flow rate and liquid level are all adjustable from the control station. These parameters are continuously monitored and recorded in a computer hard disk. In case of emergencies, such as operational failures where temperature, pressure or level exceeds preset upper limits, or when a hydrogen or H_2S leak is present, an automatic fail-safe system initiates an emergency shutdown.

Operating Conditions: The operating conditions were not the same as those of the commercial unit. We operated at conditions that gave us the same 524°C+ conversion as the commercial unit (65%). We also explored the following ranges: Temperature 400-410°C; Pressure <3000 psig; and LHSV <1h^{-1}

RESULTS AND DISCUSSION

Catalyst De-edging: Previous studies carried out in our laboratory [18] using the same feedstock indicated that it took about 120 hours for catalyst activity to stabilize. This is the relatively low catalyst age at which coking reached an equilibrium level as described by Newson [13, 14] and Dautzenberg [15]. The variations in density, viscosity, hydrodesulfurization (HDS), hydrodemetallation (HDM), hydrodenitrogenation (HDN) and Conradson Carbon Residue (CCR) reduction became low, smooth and linear with run time after 120 hours. Results for the performance of the catalysts after 120 hours of run time, based on liquid product properties, are shown in Tables 2 and 3 and illustrated in Figures 2 and 3.

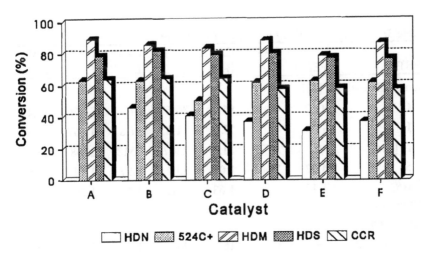

Figure 2: *Conversions for Catalysts After 120 Hour Runs*

Table 2: *Feed and Total Liquid Product Properties*

TEST	FEED	CAT. A	CAT. B	CAT. C	CAT. D	CAT. E	CAT. F
Viscosity @ 40°C [cP]	--	27.4	27.6	73.3	35.1	30.55	37.9
Viscosity @ 100°C [cP]	980	5.56	4.66	9.85	5.34	5.34	5.69
Density @ 15.5°C [g/cm^3]	1.018	0.9209	0.9179	0.9328	0.9238	0.9258	0.9278
CCR [wt%]	16	5.95	5.72	5.64	6.82	6.75	6.84
Simulated Distillation [wt%]:							
IBP-177°C	--	4.7	5.7	3.61	5.8	5.85	5.56
177-249°C	--	8.8	8.3	6.22	7.59	8.55	7.78
249-343°C	--	18.05	17.4	13.57	16.37	17.32	16.52
343-524°C	33.3	44.01	43.9	43.44	44.63	43.26	44.22
524°C+	67.7	24.12	24.7	33.17	25.61	25.03	25.93

Table 3: *Feed and Total Liquid Product Properties*

TEST	FEED	CAT. A	CAT. B	CAT. C	CAT. D	CAT. E	CAT. F
Elemental Analyses [wt]:							
C	83.7	--	86.9	86.67	87	87.66	86.48
H	10.3	--	12.3	11.87	11.8	11.67	11.56
N [ppm]	40	2472	2158	2367	2518	2768	2530
S	4.97	1.06	1.06	1.02	0.97	1.14	1.16
H/C [Atomic]	1.47	--	1.63	1.63	1.62	1.59	1.59
Metals Content:							
Ni [wppm]	79	15	16	20	16	25	20
V [wppm]	183	16	23	24	20	23	27

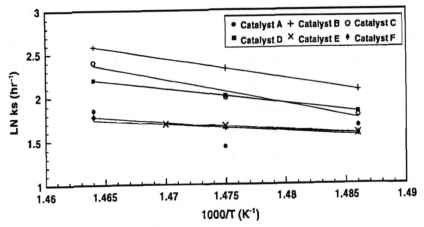

Figure 3: *HDS Activity of Catalysts*

Kinetic Studies: In order to elucidate kinetic parameters from reaction runs , reaction temperature was varied at constant pressure and LHSV. At other runs, LHSV was varied at constant reaction temperature and pressure. The Robinson-Mahoney reactor used for the runs has the advantage of approximating perfect mixing through its internal recirculation of reactants [19]. This gradientless reactor therefore permits reaction studies at isothermal conditions with uniform concentrations so that the reaction rate of each reaction can simply be calculated from an ordinary difference equation [20]:

$$\frac{W}{F} = \frac{\left(C_f - C_p\right)}{r_g} \dots\dots\dots\dots\dots\dots\dots\dots\dots\dots\dots\dots\dots\dots\dots\dots\dots\dots\dots\dots\dots\dots\dots\dots\dots\dots\dots \text{(Eq. 1)}$$

whereby , "r_g" is the global rate of reaction per unit mass of catalyst, "F" is the feed rate of reactant, "W" is the mass of catalyst , and "C" is the concentration of reactant. Subscripts "f" and "p" on the concentration refer to feed and product (concentration at the inlet and outlet). Assuming the validity of power law kinetics, i.e.:

$$r_g = k \times C_p^n \dots\dots\dots\dots\dots\dots\dots\dots\dots\dots\dots\dots\dots\dots \text{(Eq. 2)}$$

and substituting equation (2) in equation (1), we obtained the expression from where the reaction rate constant "k" was determined. The pseudo reaction order is "n" in equation (2). The reaction rate constants were calculated from runs at three different temperatures for each catalyst. The apparent activation energy and frequency factor were then determined for each catalyst and the global reaction rate was calculated based on product concentrations after 120 hours of run time. Values of apparent activation energies determined from run data for the three most active catalysts are presented in Table 4. The HDS reaction was best fitted using a pseudo reaction order of one and a half, while HDM and hydrocracking (524°C material) reactions were both fitted with pseudo first order kinetics. This is in good agreement with the work of some other researchers [15, 21, 22]. The Arrhenius plot of HDS activity of the six catalysts is shown in Figure 3.

Liquid Product Distribution: The liquid product distribution after 120 hours of run for each of the six catalysts tested is shown in Figure 4. In the yield of lighter boiling materials (IBP-249°C), catalysts B, E, D and F, showed better performance than A and C. In the yield of middle boiling distillates (249-524°C), the ranked order of performance is B>D>F>E>A>C. Pitch yield however shows another trend C>A>F>D>E>B, which is as expected exactly the opposite of the trend for lighter materials. This figure for the liquid product distribution shows that only catalyst C has conspicuously less conversion of

Table 4: *Apparent Activation Energies for HDS, HDM and Hydrocracking*

CATALYST	Activation Energy (kJ/mol)		
	HDS	HDM	HC
A	100	426	307
B	191	323	362
D	141	147	285

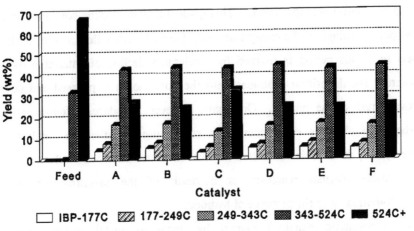

Figure 4: *Liquid Product Distribution for Hydrocracking Catalysts*

materials boiling above 343°C. Liquid products were also taken and analyzed during runs which entailed variation of reaction temperature and variation of LHSV. All results were evaluated based on HDM, HDS, CCR, HDN, 524°C conversion, and both density and viscosity reductions. Weights were applied to each factor as appropriate according to the relative importance of each of the factors to the BPU. The results indicated that catalysts A, B and D clearly outperformed the other catalysts.

Catalyst Deactivation Runs: The classical papers from Dautzenberg *et al* [15, 16] on catalyst deactivation through pore mouth plugging showed that catalyst activity loss during the initial period of a run in residue hydrocracking can be assigned to two contributions. The first is coke deposition which reaches an equilibrium level within a relatively short period of time. The second is partial surface poisoning by organometallics, specifically, mixed vanadium and nickel sulfides. We established this period to be 120 hours in our CSTR runs. The second stage in the deactivation cycle is a period when there is a steady and continuous accumulation of metal sulfides leading to a gradual catalyst activity decline. This gradual activity decline continues until the pore openings of the catalyst are fully blocked. After this, a third period sets in. This period is marked by accelerated decline of catalyst activity whereby the deposition continues until the pore mouth is plugged and coke covers most of the catalyst. Our study, however, introduces a fourth period when one observes some conversion still going on after the catalyst is presumably dead. This conversion we attribute to two main contributors:

a) Purely thermal reactions as a result of the pressure/temperature combination in the presence of hydrogen;

b) Autocatalytic reactions caused by deposited metal sulfides which themselves could act as catalysts.

Item a) was observed in our laboratory when residue hydrocracking was carried out both in the presence and in the absence of a catalyst at the same operating conditions. Conversions of metals, sulfur, CCR and pitch (524°C+) were observed [22]. The results indicated that in a no catalyst environment thermal cracking occurred with CCR conversion suffering the most and 524°C conversion suffering the least (when the conversions with and without a catalyst were compared: metals, sulfur, CCR and 524°C).

Item b) was discussed in reference [15] and the evidence for the autocatalytic reaction was shown in a figure in both references [15] and [17].

The periods discussed above are shown schematically in Figure 5 where the abbreviations are the same as those used by the Shell Research Team [15, 17], and are interpreted as follows: TSSD = Time of Steady-State Decline of catalyst activity, TSAD = Time of Start of Accelerated Decline of catalyst activity, UCL = Useful Catalyst Life is defined as the period from TSSD to TSAD, T_{min} is the minimum catalyst life, while T_{max} is the maximum catalyst life defined by TSAD.

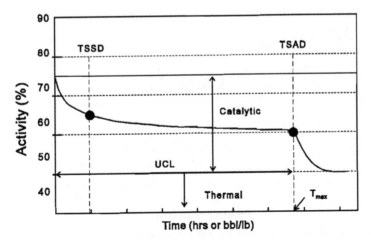

Figure 5: *Catalyst Deactivation in Hydrocracking using a CSTR*

One could easily infer from Figure 5 that the difference between the activity level of the catalyst at the initial stage of the run and that at the fourth period (i.e., little or no catalytic activity) would represent the activity contributed by the catalyst during its useful life. The remaining activity is ascribed to mainly thermal reactions. Also for design purposes, the slope of the line from TSSD to TSAD can easily be recalculated to give the increase in reaction temperature required in commercial operations in order to maintain a given conversion.

In order to ascertain when the three catalysts chosen from short term runs would be out of service, we decided to carry out reaction runs with each of the catalysts at the same operating conditions for as long as the unit was operable in order to determine their Useful Catalyst Life (UCL). The metals, sulfur, CCR and 524°C conversion profiles presented in Figures 6, 7, 8 and 9 respectively show that Catalyst A has a UCL of around 1250 hours.

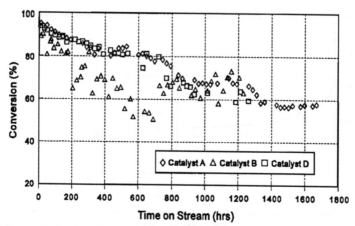

Figure 6: *Metals Conversion vs. Time on Stream*

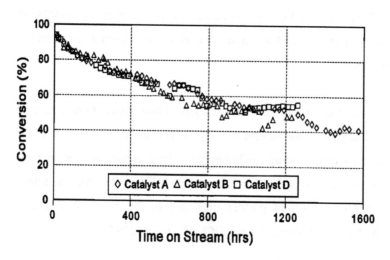

Figure 7: *Sulfur Conversion vs. Time on Stream for Long Term Runs*

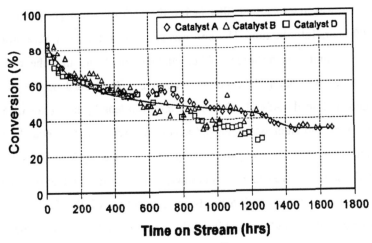

Figure 8: *CCR Conversion vs. Time on Stream*

For catalyst D, it was around 1000 hours. Catalyst B had an undefined UCL. Runs with Catalysts A and D went smoothly throughout the run period. For those two catalysts there were no shutdowns for any reason. We ran until the unit became inoperable as a result of coke build up in the reactor and the stirrer could not turn. Catalyst B, on the other hand, was prone to sediment formation

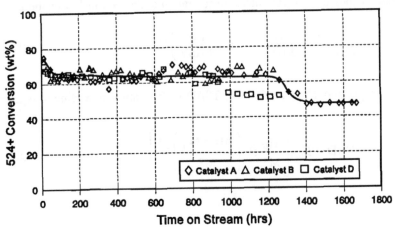

Figure 9: *524+ Conversion vs. Time on Stream*

so that the overflow tube and filter before the letdown valve got plugged up after 700 hours into the run. The unit was shut down, the lines were cleaned out and the unit was restarted. After another 100 hours into the run the overhead line was plugged by gummy deposits. We cleaned up the system again and continued. After 1230 hours of total run length, the let down valve, overflow tube and filter were plugged again and we decided to end the run. The production of sludge in hydrocracking was discussed by Van Driesen [23] and shall not be reviewed here. It is however important to note that the amount of sludge material produced depends on the nature of the feedstock, the operating conditions and the activity of the catalyst. With all other conditions being the same for all our runs, it is only logical to attribute the high sludge formation of Catalyst B to the nature of the catalyst. It is obvious that this catalyst would have more tendency to cause operational problems than Catalyst A or D.

A computer program was written which calculates the following based on reference [15]: Theta = the relative catalyst age or the degree of deactivation of the catalyst; T_{min} = the minimum catalyst life, i.e., the life of a catalyst exposed to the full metal concentration of the feedstock (Equation. 3).

$$\frac{t}{T_{min}} = \left(1 + \frac{k_M}{LHSV}\right)\theta - \left(\frac{1}{2}\frac{k_M}{LHSV}\right)\theta^2 \quad \text{.............................. (Eq. 3)}$$

Once T_{min} and Theta are known, TSAD is determined according to equation 4.

$$TSAD = T_{min}\left[\frac{M_F}{M_p^o}\theta - \left(\frac{M_F - M_p^o}{M_p^o}\right)\frac{\theta^2}{2}\right] \quad \text{.............................. (Eq. 4)}$$

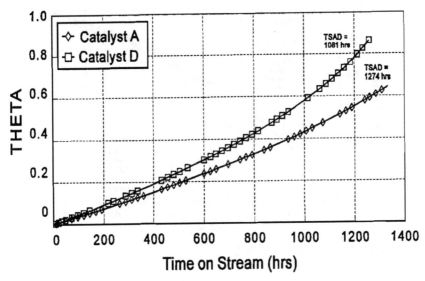

Figure 10: *THETA vs. Time on Stream*

Figure 10 shows a diagram of Theta versus catalyst age in terms of time-on-stream in hours. Catalyst age is sometimes expressed as barrels of oil processed per pound of catalyst (bbl/lb) or kilograms of oil processed per kilogram of catalyst (kg/kg). The diagram shows that the level of accumulations on catalyst D (degree of deactivation) is higher than that of catalyst A hence the calculated TSAD is higher for catalyst A than D. The calculated values are almost the same as those which could be estimated from Figures 6 to 9. Figure 11 shows the selectivity of the catalysts towards sulfur or metals removal. The figure indicates that both catalysts have a higher inclination for metals than for sulfur removal. Correlations shown would enable us to estimate the percentage of metal removal for a given percentage of sulfur removal. Figures 12 and 13 show $\ln(k_M)$ and $\ln(k_S)$ respectively versus time-on-stream for both catalysts. These figures confirm our observations from Figure 11 that catalyst A is slightly more active for metals removal but

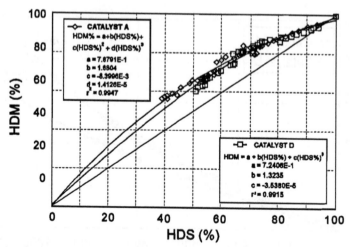

Figure 11: *Catalyst Selectivity*

deactivates at a lower rate. The desulfurization rate constant for both catalysts seem to be the same for the 1000 hours of the run. Any deviation after this time is not very relevant because the TSAD of catalyst D is 1081 hours.

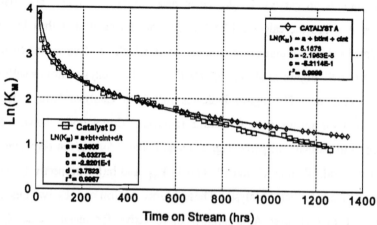

Figure 12: $Ln(K_M)$ *vs. Time on Stream*

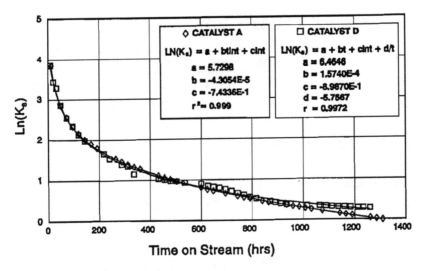

Figure 13: *Ln(KS) vs. Time on Stream*

CONCLUSION

The choice of a best catalyst for hydrocracking heavy oil or residue is a long, expensive and arduous task. The catalyst pore diameter required for optimum desulfurization is normally lower than that required for optimum demetallation. While a lower surface area with a larger pore volume is generally preferable, the surface area and pore volume of the catalysts do not tell the whole story. Initial catalyst activity as premises for catalyst choice may also be deceptive. We have carried out long term deactivation runs on three out of six originally chosen commercial catalysts. The three chosen catalysts were outstandingly superior in their performance over the other three during short term runs.

Long term runs carried out with the three catalysts showed that one of them was prone to causing sediment formation in the process equipment at our operating conditions. This catalyst was therefore eliminated for now. Of the other two catalysts, one had a useful catalyst life of 1081 hours and the other

Table 5: *Summary of Results for the Tested Catalysts*

	A		B		C	
	SOR	EOR	SOR	EOR	SOR	EO
API Gravity (API)	24.99	15.21	26.28	18.87	25.83	17.
C	--	--	87.0	85.6	86.7	85.
H	--	--	12.5	11.5	13.0	11.
H/C Ratio (Atomic)	--	--	1.71	1.60	1.78	1.6
Conversions:						
Total Metals	95.15	57.38	89.13	62.33	93.63	58.
Nitrogen	75.06	32.72	77.98	25.90	70.88	18.
Sulfur	93.79	48.86	93.71	48.16	93.86	55.
(524°C+)	74.92	47.50	72.93	66.67	72.79	52.
MCR (CCR)	82.59	34.12	82.19	41.56	82.75	2.8

1274 hours. Minimum and maximum catalyst life elucidated from run data were easily recalculated to minimum and maximum catalyst replacement rates for an ebullated-bed reaction unit. Reaction rate constants for metals and sulfur removal were determined. Catalyst characterization work carried out on both the fresh and spent catalysts did not correlate with performance data. Catalyst A proved to be a better choice than its counterpart finalist - Catalyst D based on the deactivation curve, maximum lifetime, catalyst replacement rate and overall performance. Catalyst B needs revisiting because it shows potential for higher activity (see Table 5). Most probably, if operated at less severe conditions, this catalyst might exhibit less sludge formation tendency. Other methods of reducing sludge formation as described in reference 23 may also apply.

ACKNOWLEDGMENTS

We would like to thank the management of Husky Oil Ltd., Calgary, the Bi-Provincial Upgrader, Lloydminster, and Novacor Research & Technology Corporation (NRTC), Calgary, for funding this project and allowing us to publish the results. Our gratitude goes to Leon Neumann, and the Analytical Group at NRTC for analyzing most of the samples.

REFERENCES

1. Oballa, C.M., Wong, C., Krzywicki, A.; "Proceedings of the International Symposium on Heavy Oil and Residue Upgrading and Utilization", Han Chongren and Hsi Chu, Ed.; International Academic Publishers, Beijing, pg. 133, (1992).

2. Rajagopalan, K. and Luss, D.; Ind. Eng. Chem., Proc. Des. Dev., 18, 459-465, (1979).

3. Sahimi, M. and Tsotsis, T.T.; J. Catal., 96, 552-562, (1985).

4. Oyekunle, L.O. and Hughes, R.; Chem. Eng. Res. Des., 62, 339-343, (1984).

5. Khang, S.J. and Mosby, J.F.; Ind. Eng. Chem., Proc. Des. Dev., 25, 437-442, (1986).

6. Haynes, H.W., Jr. and Leung, K.; Chem. Eng. Commun., 23, 161-179, (1983).

7. Ahn, B. and Smith, J.M.; AIChE J., 30, 739-746, (1984).

8. Beeckman, J.W. and Froment, G.F.; Ind. Eng. Chem. Fundam., 18, 245-256, (1979).

9. Beeckman, J.W. and Froment, G.F.; Chem. Eng. Sci., 35, 805-815, (1980).

10. Tsakalis, K.S., Tsotsis, T.T. and Stiegel, G.J.; J. Catal., 188-202, (1984).

11. Hiemenz, W.; Sixth World Petroleum Congress, Frankfurt, Discussion Section 3, Paper 20, June 21, (1963).

12. Beuther, H., Schmid, B.; Sixth World Petroleum Congress Frankfurt, Discussions Paper No. 20, pgs. 297-307, (1963).

13. Newson, E.; ACS Preprints, Divsision of Fuel Chemistry, Vol. 17, No. 2, pgs. 49-63, (1972).

14. Newson, E.; Ind. Eng. Chem., Proc. Des. and Dev., Vol. 14, No. 1, pgs. 27-33, (1975).

15. Dautzenberg, F.M., Van Klinken, J., Pronk, M.A., Sie, S.T., and Wiffles, J.B.; Paper presented at the 5th International Symposium on Chemical Reaction Engineering, Houston, March 13-15, (1978).

16.   Dautzenberg, F.M., George, S.E., Ouwerkerk, C., and Sie, S.T.; Paper presented at the Advances in Catalytic Chemistry II Symposium, Salt Lake City, Utah, U.S.A.; May 18-21, 1982.

17.   Sie, S.T.; in "Catalyst Deactivation", edited by Delmon, B. and Froment, G.F., published by Elsevier, Amsterdam, pgs. 545-569, (1980).

18.   Oballa, C.M., Smith, K., Lewkowicz, L., Krzywicki, A., and Wong, C.; "Control of Catalyst Deactivation in Catalytic Hydroprocessing of Bitumen and Heavy Oil Residue", Final Report to Energy, Mines and Resources Canada, Contract No. 23440-9-9287/01-SQ, Oct. 30, 1991.

19.   Mahoney, J.A., Robinson, K.K.; Paper presented at the ACS Division of Petroleum Chemistry Meeting, Chicago, Aug. 28-Sept. 2, (1977).

20.   Mahoney, J.A., Robinson, K.K.; and Meyers, E.C.; Chem. Tech., Vol. 8, pgs. 758-763, Dec., (1978).

21.   Silbernagel, B.G., and Riley, K.L.; in "Catalyst Deactivation", edited by Delmon, B. and Froment, G.F., published by Elsevier, Amsterdam, pgs. 313-321, (1980).

22.   Oballa, C.M., Wong. W.M. and Kryzwicki, A.; Paper presented at the American Institute of Chemical Engineers Annual Meeting, San Francisco, Nov. 5-10, (1989).

23.   Van Driesen, R.P. et al; Energy Processing Canada, pg. 13-19, July/August, (1987).

3   Resid Hydrocracking: New Frontiers

G. Nongbri, G. A. Clausen, J. R. Huang,
D. E. Self, C. A. Paul and A. I. Rodarte

4545 Savannah Avenue @ HWY 73
Port Authur, TX 76640

ABSTRACT

Texaco Research and Development at Port Arthur, Texas recently completed a very successful H-Oil[®] pilot plant program demonstrating the use of a Texaco/American Cyanamid newly developed catalyst for hydrocracking residual feedstocks at high conversion and still produced stable fuel oils. The operation was carried out in a Process Development Unit wherein catalyst was added and withdrawn daily. Stable fuel oil was produced at 538° C+ (1000° F+) conversion in excess of 80 volume percent. The gas oil and overhead products were treated in an in-line ebullated bed unit (T-STAR[SM]) which produced very clean distillate products (diesel <0.03 wt% sulfur and heavy gas oil <0.15 wt% sulfur).

INTRODUCTION

Even though the demand for petroleum products has decreased because of the major oil price escalations in 1973/74 and 1979/80 and periods of economic

Nongbri et al.

recession and stagnation, indications are, that after 1992, world oil consumption is expected to increase. Total petroleum products demand is expected to rise by over 1.2 million barrels per day (BPD) annually from 1992-2002.

Historical records depicted in Figure 1 show that the API gravity of the crude supply has decreased and the sulfur content has increased. These trends are predicted to continue. The decrease in API gravity and increase in sulfur are associated with an increase in residual oil content of the crude. Table I summarizes the historical change and projected future change in vacuum residua quality. Environmental considerations indicate that the sulfur content of product will decreased and fuels will have fewer olefins and aromatics.

As the refiners increase the proportion of heavier, poorer quality crude in their feedstocks, the need grows for effective processing methods to treat the fractions containing increasingly higher levels of sulfur, metals and conradson carbon residue (CCR).

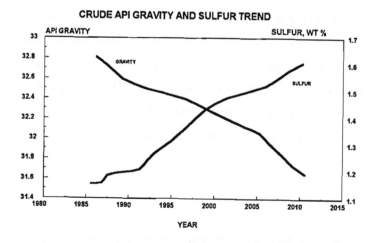

Figure 1: *Crude API Gravity and Sulfur Trend*

Table I: *World Crude Supply*

YEAR	1990	1995	2000	2005
CRUDE OIL				
GRAVITY, API	32.60	32.43	32.22	32.06
SULFUR, W%	1.20	1.31	1.44	1.52
VACUUM RESID				
YIELD ON CRUDE, V%	19.4	19.8	20.2	20.5
GRAVITY, API	9.5	9.2	9.0	8.8
SULFUR, W%	3.26	3.61	3.91	4.15
METALS, WPPM	286	297	309	319

RESID UPGRADING:

One of the objectives of resid upgrading is to maximize high-quality liquid yields. This objective can be achieved via hydrogen addition processes. One of the commercially proven hydrogen addition processes is the H-Oil Process licensed by Texaco Development Corporation (TDC) and HRI, Inc.[1]. Cokers, by comparison, yield products of lower quality which must be further hydrotreated to meet today's specifications. Since initial start-up in 1984, the H-Oil unit at the Star Enterprise refinery in Convent, Louisiana has processed virgin residua and visbroken residua obtained from a variety of crudes including Louisiana Sweet, Alaskan, Maya and Saudi Arabia.

One of the difficulties in resid or heavy oil processing is the formation of insoluble carbonaceous substances known as sediment. These substances, unless properly controlled, cause operability problems in the hydrotreater/hydrocracker and the other downstream units. Certain residua tend to produce greater amounts of solids thereby limiting the level of upgrading of these resids.

The formation of sediment also depends on the level of conversion. The higher the conversion level for a given feedstock, the greater is the amount of solids formed. Figure 2 compares the solid formation at given conversion

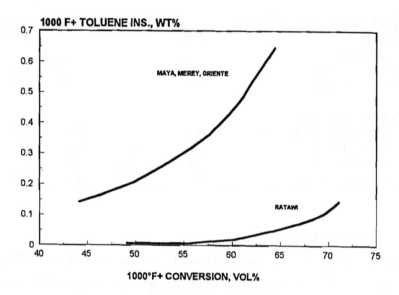

Figure 2: *Toluene Insolubles vs. Conversion*

levels of different feedstocks. With Ratawi vacuum residuum, a 538°C+ (1000° F+) conversion as high as 70 volume percent can be obtained with sediment much lower than that of the other groups of feeds. Fixed-bed VRDS processes are limited to resid conversion levels of 45-50 volume percent by sediment formation.

The formation of these solids results in the agglomeration of the catalyst, thereby causing high pressure drop, especially in fixed bed reactors. In the refinery where the hydrocracked bottoms is sold as fuel oil, the solids end up in the fuel oil and increase the sediment content of the fuel oil.

In normal operation on Arabian feeds, with the existing catalyst, the level of conversion producing stable fuel oil has been limited to less than 75 volume percent. For high sediment feedstocks, refiners have had to operate their resid conversion units at much lower conversion levels.

CATALYST DEVELOPMENT

Texaco has been active in catalyst research and in bottoms upgrading processing for several years. In a joint development with American Cyanamid (Cytec), several new catalysts were developed that performed special functions in resid upgrading operations. One of these new catalysts reduces sulfur and nitrogen in cracked distillates. Another catalyst produces low sulfur, low conradson carbon residue, high API gravity vacuum bottoms product[2]. Still other catalysts provide more $538^{\circ}C+$ ($1000^{\circ}F+$) conversion.

A new catalyst was recently developed that can process heavier, poorer quality crudes. The refiner will need to convert more of the bottoms from these crudes and, at the same time, produce stable low sulfur fuel oil. The newly developed catalyst is tailored to handle the need of the refinery so that it can

TABLE II

	FEEDSTOCK PROPERTIES
VACUUM RESID, V%	85.5
HCGO, V%	14.5
GRAVITY, API	3.2
SULFUR, W%	5.51
CARBON, W%	84.35
HYDROGEN, W%	9.68
NITROGEN, WPPM	4356
METALS, WPPM	
NICKEL	37
VANADIUM	120
CCR, W%	20.2
ASPHALTENES, W%	10.5
$538^{\circ} C^{+}$, V%	76.1
CALCULATED 538° C^{+} PROPERTIES	
GRAVITY, API	2.5
SULFUR, W%	6.08
CCR, W%	23.8

use low value crudes. The new catalyst reduces sediment at high conversion
level and provides better desulfurization than other commercially available
catalysts. This paper summarizes some of the pertinent results that are
obtained when using this catalyst.

CATALYST TESTING UNIT

The testing was carried out in a two stage H-Oil reactor Process Development
Unit (PDU). Figure 3 shows a schematic flow diagram of the PDU. The
operation of this unit was similar to the operation of the commercial unit. The
unit has atmospheric and vacuum fractionators so that recycle streams are
produced on a continuous basis. The unit also has catalyst addition and
withdrawal equipment so that catalyst can be added and withdrawn daily as
done in commercial operations. The level of catalyst in the reactor is detected

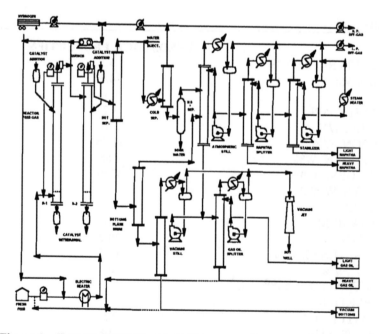

Figure 3: *F-10 H-Oil PDU with Full Recovery Section*

by a travelling gamma ray detector (density difference). Once catalyst activity reaches equilibrium, constant product yields and qualities are produced. The feed stock used in htis study was a blend of virgin Arabian medium vacuum bottoms, visbroken bottoms and heavy cycle gas oil. test results on hte feedstock are given in Table II. The feed blend sulfur content was 5.5 wt% and 76.1 volume percent of the feed boiled above 538°C+. The 538°C material had a 2.5 API Gravity with a sulfur ocntent of 6.08 wt%.

H-OIL REACTOR:

Figure 4 shows the schematic of the H-Oil (ebullated bed) reactor. The design and operation of a T-STAR reactor are similar to those of the H-Oil reactor.

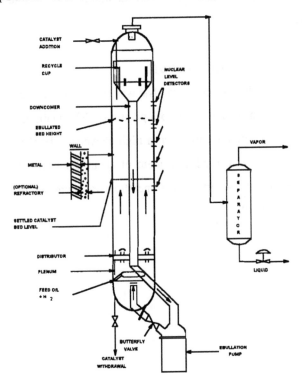

FIGURE 4:   *H-Oil Reactor*

Gas, liquid and solid catalyst are intimately mixed in the reactor. The catalyst in the reactor is kept in the fluidized state by a controlled flow of the reactor fluid. The upward flow of gas and liquid expands the catalyst bed and distributes liquid, gas and catalyst evenly across the reactor, maintaining the suspended catalyst particles in random motion. Catalyst is withdrawn and added daily during operation and thereby eliminates the need to shutdown the unit and replace deactivated catalyst. The possibility of plugging or blockage from build-up of coke or sediment in the system is virtually eliminated. The reactor operates with a low and constant pressure drop. Because of the internal recycle flow, the reactor operates at near isothermal conditions. A detailed description of the reactor, important design considerations, and advantages are given in the reference[3].

CATALYST EVALUATION IN THE PDU:

The evaluation of the newly developed catalyst was carried out in the PDU for 110 days of operation with several modes of operations. The main objective of the operation was to evaluate the catalyst at a high conversion level and while producing stable fuel oil with a sulfur content of about 2.0 wt%. Once the catalyst activity lined out, the severity of the operation was increased in order to increase the conversion level.

Figure 5 summarizes the sediment numbers obtained on the bottoms flash drum (BFD) product using the IP-375 test method. This stream is in composition to the atmospheric bottoms stream in the commercial unit. The BFD sediment (wt%) is plotted against 538°C + conversion at conversion levels ranging from 56 to 82 volume percent conversion. The sediment numbers obtained with the new catalyst are quite low (<0.07 wt%) and did not show any increase even as conversion was increased. At the 55-70 volume percent conversion level, the sediment numbers obtained with the new catalyst are 0.09 to 0.14 wt% lower than those obtained with the standard ebullated bed catalyst.

It should be noted that in all of the comparisons of the new catalyst versus the standard catalyst in these studies, the liquid space velocity and catalyst addition rate were similar.

Vacuum bottoms obtained from this operation at 66, 74 and 81 volume percent 538°C+ conversion were blended with hydrotreated light cycle gas oil to a concentration of 30 wt% to produce fuel oil. The results are summarized in Table III. The sediment contents of the fuel oils were very low at all

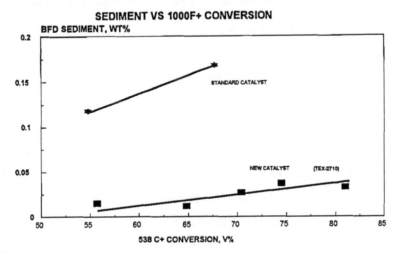

Figure 5: *Sediment vs 1000°F + Conversion*

TABLE III: *Fuel Oil blending Results*

$538^{0}C^{+}$ CONVERSION, V%	66	74	81
H-OIL BOTTOMS, W%	70	70	70
HIGH SULFUR CUTTER STOCK, W%	30	30	30
FUEL OIL PROPERTIES			
GRAVITY, API	10.2	7.7	5.3
SULFUR, W%	2.08	2.10	2.10
CCR, W%	18.6	20.6	27.1
VISCOSITY, cSt @ 50^{0} C	273	339	539
EXISTENT SEDIMENT, W%	0.02	0.00	0.03
POTENTIAL SEDIMENT, W%	0.01	<0.01	0.03

conversion levels. These results indicate that very stable fuel oil was produced at conversions greater than that normally obtained with the standard catalyst.

For the operation at around 85 volume percent conversion, the amount of H-Oil bottoms was sufficient to generate enough hydrogen to balance the process hydrogen consumption. With the standard catalyst, operation between the conversion range of 75 to 85 volume percent had previously presented a "no man's land" situation since there was too much sediment in the product to allow use of the unconverted bottoms in heavy fuel oil and there was too much bottoms left for what was usually needed for hydrogen generation. Disposal of this excess material thus became aproblem. The new catalyst allowed production of stable fuel oil in this conversion range and thus provided the refiner with flexibility to produce products that are in demand at different times.

Comparative results on the fuel oil blend obtained from the operation with the new catalyst to that obtained on the standard catalyst is shown in Table IV. The results show that the new catalyst produced a more stable fuel oil than the standard catalyst as indicated by the low potential sediment analysis (0.05 wt% for the new catalyst as compared to 0.20 wt% for the standard catalyst).

For the same amount of cutter stock, the new catalyst gave a lower sulfur fuel oil (1.84 versus 2.08 wt%). Figure 6 summarizes the $538^{\circ}C+$ ($1000^{\circ}F+$) sulfur of the product as the conversion was increased. Comparative sulfurs when using standard catalyst are also presented. The new catalyst produces a $538^{\circ}C+$($1000^{\circ}F+$) bottoms product with about 0.3 wt% lower sulfur than the standard catalyst. The new catalyst produced a $538^{\circ}C+$ product with about 2 wt percent lower conradson carbon and about 2.5 to 3 API Gravity higher than the standard catalyst. The lower conradson carbon material, when charged to the coker, will result in improved liquid yields and lower the coke yield from the coker.

TABLE IV: *Comparison of Fuel Oil Blends*

CATALYST	STD	TEX-2710
538° C^{+} CONVERSION, V%	54	56
H-OIL BOTTOMS, W%	70	70
LOW SULFUR CUTTER, W%	30	30
FUEL OIL PROPERTIES		
GRAVITY, API	10.5	12.3
SULFUR, W%	2.08	1.84
CCR, W%	NA	16.9
VISCOSITY, cSt @ 50° C	220	237
EXISTENT SEDIMENT, W%	0.19	0.02
POTENTIAL SEDIMENT, W%	0.20	0.05

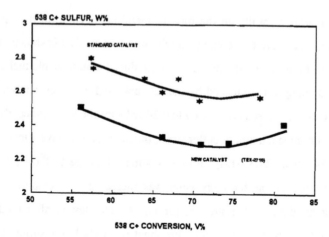

Figure 6: *538 C+ Sulfur vs. Conversion*

Figure 7 compares the sulfur results of the heavy gas oil. As in the case of the bottoms sulfur, the new catalyst produced heavy gas oil with about 0.15 wt% lower sulfur than that obtained with the standard catalyst. Also, the corresponding nitrogen content of the heavy gas oil is lower by 300 wppm.

The use of this catalyst provides the refiner with the flexibility of

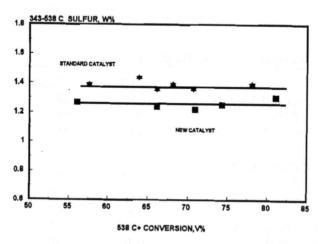

Figure 7: *Vacuum Gas Oil Sulfur vs. Conversion*

operating his H-Oil unit to take advantage of market conditions. If there is a market demand for fuel oil, he can run the H-Oil unit at low conversion and maximize fuel oil production. If there is a higher demand for distillate products, the H-Oil unit can be operated in a high conversion mode to maximize distillate production and at the same time produce stable lower sulfur fuel oil.

It should be noted that for either catalyst, the sulfur in the products can be further lowered by decreasing the liquid space velocity and/or increasing the fresh catalyst addition rate.

COMMERCIALIZATION OF THE CATALYST

As a result of the outstanding performance of the new catalyst in the PDU, half a million pounds of this catalyst were manufactured by Criterion. The catalyst was successfully tested in the Star Enterprise H-Oil unit in Convent, Louisiana with resid conversion as high as 84 volume percent.

IN-LINE HYDROTREATING

In order to provide very high quality distillate product for use as blending stock with other lower quality distillates, experiments were conducted wherein the gaseous effluent from the H-Oil unit was fed without de-pressuring, to another ebullated bed unit (T-STAR). All distillate products, except the resid, are processed in this T-STAR unit to remove sulfur and nitrogen. The catalyst used in the distillate unit is the same as the one used in the H-Oil unit so that the spent catalyst from the distillate unit can be cascaded to the H-Oil reactors. The T-STAR reactor can be operated either as a hydrotreater to remove sulfur, nitrogen, aromatics, etc. or as a mild hydrocracker to further increase distillate production.

TABLE V: *Combined Performance from H-Oil/In-Line T-Star Operation*

538°C+ CONVERSION, V%	81
	93.2
DENITROGENATION, W%	73.8
VANADIUM REMOVAL, W%	97.2
NICKEL REMOVAL, W%	82.8
CCR REMOVAL, W%	67.2
CHEMICAL HYDROGEN	
CONSUMPTION, Nm^3/m^3	352
NO OF STAGES H-OIL & T-STAR	3

Table V, VI, and VII summarizes results obtained from a combined operation. In this operation, the vacuum residuum was hydrocracked to 81 volume percent 538°C+ conversion in the H-Oil unit and the distillate was hydrocracked to 40 volume percent 343°C+ conversion in the T-STAR unit. With this operation, diesel that was below the sulfur specification and a vacuum gas oil that would be an excellent FCCU feed were produced.

Inclusion of an in-line T-STAR hydrotreater/hydrocracker in an H-Oil unit system produces high quality distillates meeting environmental regulations expected in the future.

TABLE VI: *Combined Yields From H-Oil/In-Linw T-Star Operation*

FRACTIONS	WT%	VOL%
H_2S & NH_3	5.84	
C_1	1.56	
C_2	1.68	
C_3	2.01	
C_4	1.62	2.93
C_5 - 82C	2.09	3.37
82 - 182C	13.05	17.77
182 - 343C	37.08	45.34
343 - 538C	22.92	26.52
538C+	15.15	14.26
TOTAL	103.00	110.19

TABLE VII: *Combined H-Oil/In-Line T-Star Operation Product Properties*

FRACTIONS	C_5 - 82 C	82 - 182 C	182 - 343C	342 - 538 C	538 °C+
GRAVITY, API	85.0	51.8	33.2	24.3	-4.8
SULFUR, W%	0.001	0.004	0.016	0.052	2.37
NIT, WPPM	1	5	22	336	6964
CCR, W%					43.7
V, WPPM					22
Ni, WPPM					42

CONCLUSIONS

A new H-Oil catalyst has been developed by Texaco/American Cyanamid for hydrocracking vacuum residua which will allow stable fuel oil production at $538^{0}C+$ conversions in excess of 80 volume percent.

ACKNOWLEDGEMENTS:

We would like to express our thanks to Charles H. Schrader, William B. Livingston, David P. Arceneaux, Harold C. Kaufman and Avilino Sequeira for providing comments and information during the preparation of this paper.

LITERATURE CITED

1. Tasker, K. G., L. I. Wisdom, W. B. Livingston and S. M. Sayles, "Texaco H-Oil Unit Commercial Operations," Japan Petroleum Institute Petroleum Refining Conference, Tokyo (1988).

2. Dai, E. P., D. E. Sherwood, Jr. and B. R. Martin, "Effect of Diffusion On Resid Hydrosulfurization Activity," Chemical Engineering Science 45, 8 (1990):2625-2629.

3. Nongbri, G and G. A. Clausen "Commercial Application of The Ebullated Bed Technology," **FLUIDIZATION VII** - Proceedings of the Seventh Engineering Foundation Conference on Fluidization, Brisbane, Australia (May 3-8, 1992).

4 Upgrading of a Moroccan Deasphalted Shale Oil over Mechanical Mixtures of Sulfided Cobalt–Molybdenum and Nickel–Molybdenum Alumina Supported Catalysts

Claude Moreau[1], Abdennaji Benyamna[2], Chakib Bennouna[2] and Patrick Geneste[1]

[1] Laboratoire de Chimie Organique Physique et Cinétique Chimique Appliquées, URA CNRS D0418, Ecole Nationale Supérieure de Chimie, 8 Rue de l'Ecole Normale, 34053 Montpellier Cedex 1, France

[2] Laboratoire de Chimie Organique Appliquée, Faculté des Sciences, Université Cadi Ayyad, BP S15, Marrakech, Morocco

ABSTRACT

Experimental factorial design was used to study the influence of the different parameters such as the reaction temperature, the hydrogen pressure and the reaction time on the hydroprocessing of a deasphalted shale oil over mechanical mixtures of sulfided cobalt-molybdenum and nickel-molybdenum alumina supported catalysts. It was shown that hydrodesulfurization, hydrodeoxygenation hydrodenitrogenation and hydrodearomatization were more important for high temperature, high pressure and long reaction time operating conditions as generally observed for separate experiments carried under conditions of industrial catalytic tests. The most striking feature was the existence of a promotion effect due to the simultaneous presence of those catalysts mechanical mixtures, i.e. cobalt-molybdenum-rich mixtures are more efficient for hydrodenitrogenation reactions, whereas nickel-molybdenum-rich mixtures exhibit a better activity for hydrodesulfurization and hydrodearomatization reactions, thus confirming first our previous findings in this field concerning the influence of cobalt and nickel promoters and then the general knowledge on the separate behavior of sulfided cobalt-molybdenum and nickel-molybdenum alumina supported catalysts.

INTRODUCTION

It has been recently shown that mechanical mixtures of sulfided cobalt-molybdenum and nickel-molybdenum alumina supported hydrotreating catalysts exhibited higher activity for the overall conversion of quinoline (1). The enhancement of the reactivity can reach a factor of 3, due mainly to the better hydrogenolysis properties of the catalysts mixtures in the nitrogen removal steps, ultimate goal in hydrodenitrogenation reactions.

In this paper, we wish to report on the activity of such mechanical mixtures in the hydroprocessing of a moroccan deasphalted shale oil, by emphasizing on the hydrogenolysis and hydrogenation steps, under different operating conditions of temperature, hydrogen pressure and reaction time.

EXPERIMENTAL

Catalysts

The alumina supported cobalt-molybdenum and nickel-molybdenum catalysts were Procatalyse HR 306 and HR 346 respectively. They were sulfided separately at atmospheric pressure with a gas mixture of 15% H_2S and 85% H_2 by volume. The catalysts (particule size 0.063- 0.125 mm) were heated in flowing H_2/H_2S (gas flow, 120 ml/min) from 20 to 400°C (8°C/min) and held at 400°C for 4 h, then cooled, and finally swept with nitrogen for 30 min.

Hydrotreating experiments

Experiments were carried out in a 0.3 litre stirred autoclave (Autoclave Engineers Magne-Drive) operating in a batch mode at desired temperature and hydrogen pressure.

Procedure

The deasphalted shale oil (60 ml) was poured into the autoclave. The mixture of separately sulfided catalysts (0.6 g) was rapidly added to this solution under nitrogen to avoid contact with air. After it had been purged with nitrogen, the temperature was increased until it reached the desired temperature, 200°C (level -) or 350°C (level +). Hydrogen was then introduced at the required pressure, 30 bar (level -) or 70 bar (level +). Zero time was taken to be when the agitation began. After 2h (level -) or 6h, (level +), the reactor was cooled and the reaction mixture analyzed.

Analyses
Elemental analyses were performed by the Service Central d'Analyse of CNRS, Vernaison, France.

Oil deasphalting
The crude shale oil was deasphalted with hexane in a classical manner (2). Hexane was added to the crude oil in the weight ratio hexane to oil 10:1. After agitation for 8h and decantation, the maltene fraction is obtained after filtration on Whatman paper (N°.2). The composition of the deasphalted oil is the following : C (79.6%), H (9.8%), N (1.4%), O (2.2%) and S (7.5%).

RESULTS AND DISCUSSION

Experimental design
Experimental factorial design aims at limiting the number of experiments normally required to study the influence of the most important factors involved in a given reaction (3). The choice of these factors thus appears to have direct consequences on the quality of the factorial design. Each factor must be represented by two different levels.

For hydrotreatment of the deasphalted shale oil, the composition of the feed is given, and the principal factors to consider are then the reaction temperature, the hydrogen pressure and the reaction time on the different series of catalysts mixtures. After preliminary experiments (2), two levels were considered to account for the influence of temperature, pressure and reaction time. These factors, reported in Table 1, were chosen on the basis of the knowledge acquired in the study of numerous series of model compounds over both cobalt-molybdenum and nickel-molybdenum sulfided catalysts (4).   .

Temperature : Hydrotreatment reactions are, of course, influenced by the temperature which acts on both kinetics and thermodynamics of the reaction. Moreover, they are not influenced in a similar manner. Sulfur removal is generally easier than nitrogen removal under low operating conditions, both heteroatoms being, on the other hand, more easily removed for saturated than for aromatic molecules.

Hydrogen pressure : The influence of hydrogen pressure mainly concerns reaction equilibria between aromatics and saturated heteroatoms-containing molecules, particularly for nitrogen-containing compounds for which high pressures are generally required, a zero order in hydrogen being obtained for pressures higher than 70 bar.

Reaction time : A short reaction time was chosen to account for the easy removal of heteroatoms and a longer one for the more difficult removal reactions.

For a comparative purpose, the higher parameters correspond to industrial operating conditions for which most of the classical aromatic models such as thiophene, benzothiophene, quinoline, indole, etc... are hydroprocessed.

Table 1 : Factors and levels of the factorial design.

Factors	Level (-)	Level (+)
F1 : Reaction temperature	200°C	350°C
F2 : Hydrogen pressure	30 bar	70 bar
F3 : Reaction time	2h	6h

The experimental responses are the percentages of hydrodesulfurization, hydrodeoxygenation, hydrodenitrogenation and the atomic ratio H/C indicative of the hydrodearomatization step. They are generally accounted for in terms of a polynomial mathematical model :

$$Y = a_0 + a_iF_i + a_{ij}F_{ij} + a_{ijk}F_{ijk}$$

in which a's are the coefficients of the main effects and interactions for the different responses and F's the different factors considered.

For 3 factors and 2 levels, 8 (2^3) experiments are required to calculate the coefficients of the main effects (a_i) and those for interactions between two (a_{ij}) or three factors (a_{ijk}). This corresponds to the matrix of experiments reported in Table 2.

Table 2 : Matrix of experiments.

Experiment N°	F1 (temperature)	F2 (pressure)	F3 (time)
1	+	+	+
2	-	+	+
3	+	-	+
4	-	-	+
5	+	+	-
6	-	+	-
7	+	-	-
8	-	-	-

Moreover, each set of experiments was carried out for the different compositions of the catalysts mixtures which are given in Table 3.

Table 3 : Composition (wt %) of the catalysts mixtures.

Catalyst N°	Catal. 1	Catal. 2	Catal. 3	Catal. 4	Catal. 5
CoMo	100%	75%	50%	25%	0%
NiMo	0%	25%	50%	75%	100%

Table 4 : Experimental factorial design and corresponding responses.

Exp N°	F1	F2	F3	% HDS	% HDO	% HDN	H/C (at)
11	+	+	+	61.73	58.18	21.32	1.62
12	-	+	+	01.73	09.77	24.26	1.46
13	+	-	+	53.47	54.54	13.23	1.59
14	-	-	+	01.80	14.55	02.21	1.51
15	+	+	-	39.47	40.00	11.03	1.60
16	-	+	-	00.00	12.91	00.07	1.64
17	+	-	-	32.67	42.27	05.88	1.55
18	-	-	-	13.30	16.82	04.41	1.48
21	+	+	+	57.60	50.00	26.47	1.68
22	-	+	+	01.20	09.54	05.15	1.42
23	+	-	+	47.07	45.45	00.07	1.63
24	-	-	+	05.33	24.54	00.00	1.47
25	+	+	-	31.20	37.27	07.35	1.64
26	-	+	-	04.40	20.45	27.94	1.52
27	+	-	-	28.13	33.41	07.35	1.58
28	-	-	-	00.00	05.45	05.15	1.47
31	+	+	+	76.00	54.09	14.71	1.61
32	-	+	+	05.07	21.36	00.00	1.54
33	+	-	+	59.07	37.73	07.35	1.55
34	-	-	+	02.80	30.91	11.03	1.52
35	+	+	-	49.60	42.73	11.03	1.58
36	-	+	-	00.00	15.45	02.94	1.50
37	+	-	-	33.27	30.68	05.88	1.62
38	-	-	-	08.27	42.27	03.68	1.51
41	+	+	+	69.33	52.73	13.23	1.86
42	-	+	+	03.87	22.27	11.76	1.49
43	+	-	+	58.73	42.95	06.62	1.56
44	-	-	+	01.33	27.73	10.29	1.51
45	+	+	-	50.27	41.36	09.56	1.64
46	-	+	-	05.47	12.73	04.41	1.53
47	+	-	-	41.60	34.54	15.44	1.57
48	-	-	-	03.07	17.73	09.07	1.49

Table 4 (continued)

Exp N°	F1	F2	F3	% HDS	% HDO	% HDN	H/C (at)
51	+	+	+	56.93	58.64	18.38	1.65
52	-	+	+	04.53	20.45	00.00	1.48
53	+	-	+	58.93	63.63	00.00	1.59
54	-	-	+	00.00	16.82	06.62	1.49
55	+	+	-	48.80	44.54	05.88	1.64
56	-	+	-	05.33	23.18	02.91	1.48
57	+	-	-	38.00	39.09	05.15	1.52
58	-	-	-	03.93	09.09	05.88	1.51

Note : For experiments numbering, the first figure corresponds to the composition of the catalysts mixtures, as given in Table 3 and the second figure corresponds to the fractional factorial design, as reported in Table 2.

Analysis of the factorial design results

The experimental factorial design and the corresponding responses expressed as the percentages of sulfur, oxygen and nitrogen atoms removal are given in Table 4, together with the atomic H/C ration representative of the degree of hydrogenation of the feed.

From Table 4 it can be seen that heteroatoms removal and hydrogenation percentages are, as it could be expected from previous results in this field, obtained for elevated operating conditions of temperature, hydrogen pressure and reaction time. These results are illustrated in Figs. 1-3 for hydrodesulfurization, hydrodeoxygenation and hydrodenitrogenation, respectively, as a function of the composition of the catalysts mixtures (circles). As it was already observed for hydrodenitrogenation of quinoline (1), the presence of catalysts mixtures leads to a promotion effect whatever the reaction considered. Cobalt-molybdenum rich mixtures are more efficient for hydrodenitrogenation, whereas nickel-molybdenum mixtures exhibit a better activity for hydrodesulfurization and hydrodearomatization steps.

The calculated values of the coefficients of the main effects and interactions for hydrodesulfurization, hydrodeoxygenation and hydrodenitrogenation as a function of the composition of the catalysts mixtures are given in Tables 5,6 and 7, respectively.

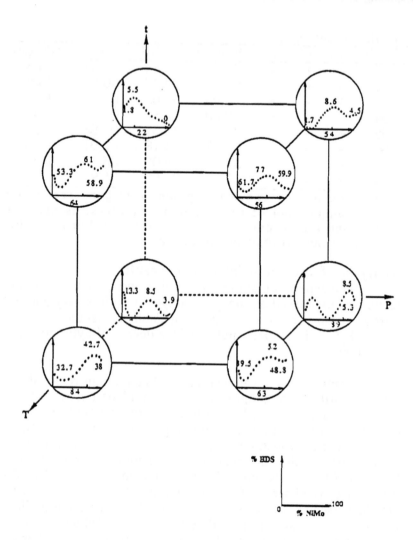

Fig.1 : Illustration of temperature, hydrogen pressure and reaction time effects, and as a function of the composition of the catalysts mixtures (circles) for hydrodesulfurization reaction of the shale oil.

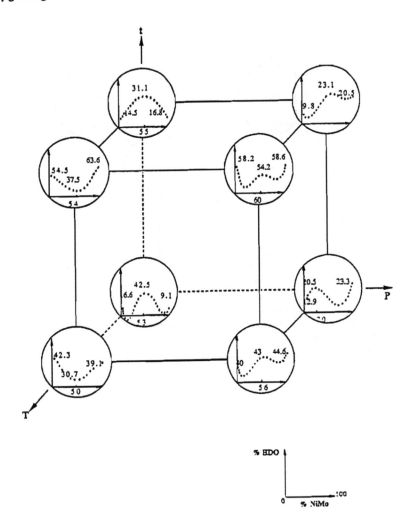

Fig.2 : Illustration of temperature, hydrogen pressure and reaction time effects, and as a function of the composition of the catalysts mixtures (circles) for hydrodeoxygenation reaction of the shale oil

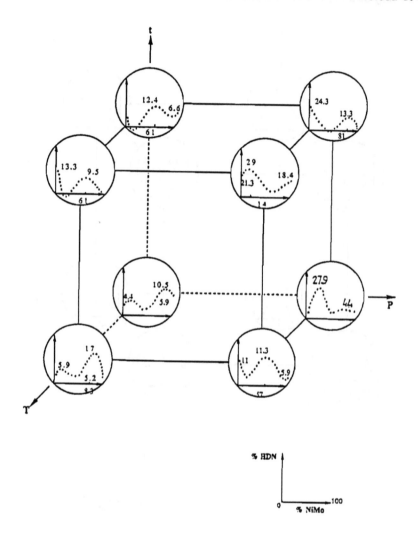

Fig.3 : Illustration of temperature, hydrogen pressure and reaction time effects, and as a function of the composition of the catalysts mixtures (circles) for hydrodenitrogenation reaction of the shale oil.

Table 5 : Coefficients of the main effects and interactions for hydrodesulfurization reaction.

Coefficient	Catal. 1	Catal. 2	Catal. 3	Catal. 4	Catal. 5
a_0	25.42	21.87	29.26	29.21	27.06
a_1	21.31	19.13	25.23	22.97	23.61
a_2	0.21	1.73	3.41	3.03	1.84
a_3	4.16	5.93	6.48	4.11	3.04
a_{12}	3.55	1.67	4.91	1.79	0.36
a_{13}	6.60	5.40	6.58	4.94	3.24
a_{23}	1.84	- 0.13	1.39	0.26	- 1.21
a_{123}	0.19	1.10	- 1.24	0.22	- 1.10

Hydrodesulfurization : As observed in Fig.1 and in Table 5, primary coefficients a_1 are higher than the others for hydrodesulfurization of the shale oil over the mechanical mixtures considered. The role of the temperature thus appears to be the most important factor on this reaction, even at low hydrogen pressure and reaction time. Although a detailed analysis of the compounds initially present in the feed was not available, this indicates that an easy removal of sulfur only by increasing temperature can take place mainly if sulfur is present in the form of lateral chains or monocyclic structures readily accessible to hydrotreating, as it was observed for distillates resulting from the distillation of the crude oil (2). The deasphalted distillates were shown to contain a higher proportion of paraffinic than aromatic compounds.

The effect of reaction time is less important by iyself but is present in the secondary coefficient a_{13} together with the factor temperature, both factors being implied on the kinetics of hydrotreating.

Hydrodeoxygenation : A similar behavior to that reported for hydrodesulfurization is also observed for hydrodeoxygenation, the major factors influencing the percentage of removal of oxygen atoms being the temperature (a_1) and to a lesser extent the reaction time (a_3), as confirmed by the secondary effect (a_{13}) involving both temperature and reaction time effects. Similar conclusions to those drawn for hydrodesulfurization can also be drawn for oxygen removal, i.e. oxygen would be present in the form of lateral chains or monocyclic structures readily accessible to hydrotreating.

Table 6 : Coefficients of the main effects and interactions for hydrodeoxygenation reaction.

Coefficient	Catal. 1	Catal. 2	Catal. 3	Catal. 4	Catal. 5
a_0	31.13	28.26	34.40	31.55	34.43
a_1	17.61	13.27	6.91	11.39	17.05
a_2	- 0.91	1.05	- 0.99	0.77	2.27
a_3	3.12	4.12	1.62	4.92	5.46
a_{12}	1.25	1.05	8.10	3.38	- 2.16
a_{13}	4.49	2.07	2.98	4.94	4.22
a_{23}	0.63	- 3.66	2.70	0.31	- 2.61
a_{123}	0.85	3.84	- 1.62	0.43	0.00

Table 7 : Coefficients of the main effects and interactions for hydrodenitrogenation reaction.

Coefficient	Catal. 1	Catal. 2	Catal. 3	Catal. 4	Catal. 5
a_0	10.30	9.94	7.08	10.05	5.60
a_1	2.56	0.38	2.67	1.17	1.75
a_2	3.87	6.79	0.09	- 0.31	1.19
a_3	4.95	2.01	1.20	0.43	1.65
a_{12}	- 0.56	- 0.19	3.03	0.49	3.59
a_{13}	- 0.54	4.97	0.09	- 1.72	1.19
a_{23}	3.67	1.10	- 1.01	2.33	1.79
a_{123}	- 2.93	5.51	1.56	0.80	2.04

Hydrodenitrogenation : For this reaction, the analysis of the primary and secondary effects is less evident. For the most active catalysts mixtures (Catal.1 and Catal.2), the main factors influencing the percentage of nitrogen removal seem to be the hydrogen pressure (a_2) and, to a lesser extent, the reaction time (a_3), both factors being involved in thermodynamic and kinetical aspects of the reaction. The higher interaction coefficients a_{23} and a_{123} are in agreement with that behavior which was not really unexpected as far as the effect of the hydrogen pressure is known to be a major one for hydrogenation of aromatics, particularly those containing nitrogen atoms. The low percentage od denitrogenation observed whatever the composition of the catalysts mixtures would mean a large amount of aromatic nitrogen compounds in the initial feed.

In order to summarize all these results, we report in Table 8 the experimental results obtained under the highest set of operating conditions which are more representative of what is generally found in the literature for hydrotreating of series of model compounds over sulfided cobalt-molybdenum and nickel-molybdenum catalysts.

Table 8 : Composition (wt %) of the catalysts mixtures and percentages of heteroatoms removal after 6h hydroprocessing of the deasphalted shale oil at 350°C, 70 bar H_2, batch reactor.

CoMo	100%	75%	50%	25%	0%
NiMo	0%	25%	50%	75%	100%
S removal	61.7%	57.6%	76.0%	69.3%	56.9%
O removal	58.2%	50.0%	54.1%	52.7%	58.6%
N removal	21.3%	26.5%	14.7%	13.2%	18.4%

As it was already observed for hydrodenitrogenation of quinoline under similar operating conditions (1), the presence of catalysts mixtures leads to a promotion effect for hydrotreating of deasphalted shale oils, whatever the reaction considered, hydrodesulfurization, hydrodeoxygenation, hydrodenitrogenation or hydrodearomatization. Cobalt-molybdenum rich mixtures are more efficient for hydrodenitrogenation, whereas nickel-molybdenum mixtures exhibit a better activity for hydrodesulfurization and hydrodearomatization steps. However, the origin of this promotion effect is not clearly understood. According to our previous results and those presented in this work, and in the absence of more precise data on the nature of the active phase, it seems that the higher hydrogenolytic properties of the cobalt-molybdenum-rich catalysts mixtures would favor a bifunctional mechanism, each part of the catalytic mixture having its own activity. Nevertheless, these assumptions have to be confirmed by other techniques and new series of experimental data.

CONCLUSIONS

Use of experimental factorial design appears to be an excellent tool to study the influence of important factors with a minimum and significant number of experiments. In this work, we have shown that a physical combination of conventional presulfided nickel-molybdenum and cobalt-molybdenum alumina supported catalysts leads to a substantial promotion effect for achieving hydrotreatment of a deasphalted shale oil, as it was already shown in separate experiments performed on model compounds, thus confirming the validity of our experimental factorial design approach. The reasons for such a promotion effect have not yet been well established. However, this effect was thought to result from either a novel active phase or through a bifunctional mechanism.

For Morocco, where shale feedstocks are very important and where sulfur content is higher than nitrogen and oxygen content with respect to other shales, upgrading of shale oils obtained by pyrolysis is an alternative and attractive route to meet the local demand for light hydrocarbons. Taking into account the nature of the feed to be hydrotreated, it appears therefore possible to find mechanical mixtures of conventional hydrotreating catalysts and their correponding optimal operating conditions to reduce operating costs.

ACKNOWLEDGEMENTS

This work was performed in the framework of an exchange agreement between the Centre National de la Recherche Scientifique (France) and the Centre National de Coordination et de Planification de la Recherche (Morocco). Oil samples were a generous gift from the Office National et d'Exploitation Pétrolière.

REFERENCES

1.   Moreau, C., Bekakra, L.,Durand, R., and Geneste, P.,
      Catalysis Today, **10** (4), 681-687, 1991.
2.   Benyamna, A., Bennouna, C., Moreau, C.,and Geneste,
      P., *Fuel*, **70** (7), 845-848, 1991.
3.   Mathieu, D., and Phan Tan Luu, R., "Méthodologie de la
      Recherche Expérimentale. Matrice d'Expérience appliquée
      aux Mélanges", Cours Université Aix-Marseille (1983).
4.   Moreau, C., and Geneste, P., "Theoretical Aspects of
      Heterogeneous Catalysis", 256-310, Ed. J.B.Moffat, Van
      Nostrand Reinhold, New-York, 1990.

5 Rapid Hydropyrolosis of Resid Oil

V.K. Mathur, M.A. Salahuddin and A.R. Mohamed

Department of Chemical Engineering

University of New Hampshire

Durham, NH 03824

In order to provide the industrialized world with sufficient
inexpensive hydrocarbon fuels and chemical feedstocks, petroleum crude has
to be utilized to its maximum. Hydrogenation of residual oil, obtained from a
petroleum distillation unit, provides a product with an increased hydrogen
content. However, to increase the hydrogen content of the product to the
level necessary for transportation fuels or chemical feedstocks, catalysts
which promote hydrogenation and hydrocracking must be employed. Catalytic
reactions play an important role in residual oil hydrogenation yielding
products of lower molecular weight with higher hydrogen to carbon ratio [1,
2]. The removal of S, N, and O heteroatoms and hydrogenation of the
residual oil molecule is usually accomplished by using commercial catalysts

containing various combinations of Co, Ni, Mo, and W on Al_2O_3 or Al_2O_3-SiO_2 supports.

Catalyst deactivation is a major problem in the catalytic residual oil hydrogenation process. Coke formation decreases catalyst activity by blocking catalyst active sites, primarily through choking the pore mouths. Deposits of nickel and vanadium deactivate the catalyst permanently by restricting diffusion paths of the reactant molecules. Even regeneration of commercial catalysts may not restore the original hydrogenation activity. Therefore, commercial supported catalysts used in residual oil hydrogenation are difficult to regenerate and have a short life. The cost of commercial catalysts is one of the major problems in this area.

Lopez et al. [3] listed the advantages of a dispersed-phase water soluble catalyst over the commercially used supported catalysts. The water soluble catalyst is a highly active form of molybdenum sulfide. The results from the hydrogenation of residual oil using a dispersed-phase catalyst show that the catalyst gives a high degree of desulfurization and demetallation (nickel and vanadium removal). The small particle size of the catalyst provides two advantages. First, the catalyst is highly active due to a large specific surface area, and second, the catalyst is sufficiently small to be readily dispersed in the residual oil allowing the oil to be easily pumped. They also claim that moderate or relatively large amounts of nickel and vanadium can be deposited on the catalyst surface without reducing its

activity. Moreover, the recycled catalyst can accommodate as much as 70 to

85 weight percent of nickel and vanadium without excessive loss of activity.

In previous studies [4, 5, 6], it has been shown that some of the ore

concentrates can be directly used for coal hydrogenation reactions. Based on

this premise, efforts have been made to use sulfide or oxide concentrate as a

catalyst or a starting point for the preparation of dispersed molybdenum

catalysts for residual oil hydrogenation reactions. It is obvious that this

procedure will result in cost reduction because of a significant saving in

material and processing expenses. A study of hydrogenation of residual oil

using a dispersed water soluble ammonium molybdate catalyst was conducted

by Mohamed and Mathur [7]. In this study, hydrogenation of atmospheric

residual oil was carried out in the presence of a dispersed water soluble

ammonium molybdate catalyst prepared from a molybdenum ore. This

provided an effective economic alternative to the use of expensive

commercial catalysts for the hydrogenation of residual oil. An ASTM

distillation of the hydrogenated residual product gave 30 percent liquid

boiling below 215 °C. Rapid hydropyrolysis of coal using various heating

techniques has been studied by several workers [8 - 15]. However, the rapid

hydropyrolysis of atmospheric or vacuum residual oil has not been reported

in the literature.

An alternative for hydrogenation of residual oil is to conduct the

process without the use of a catalyst. In this study, a novel technique known

as rapid hydropyrolysis is investigated as a way to hydrogenate residual oil.

The advantages of this technique are the absence of a catalyst and high

pressure equipment. In this investigation, a high intensity light beam is used

as a radiative heat source. The use of a concentrated solar beam or high

temperature emitter burner as a potential heat source for commercial plants is

discussed elsewhere [16].

EXPERIMENTAL EQUIPMENT AND PROCEDURE

Rapid resid oil hydropyrolysis experiments are conducted in a batch

system. The major materials used consist of atmospheric and vacuum resid

oil, hydrogen and helium gases, acetone, and tetrahydrofuran. Analyses of

resid oils are presented in Table I.

Table I

Analysis of Arabian Light Atmospheric Residual Oil and Vacuum Residual Oil

Percentage	Atmospheric	Vacuum
Carbon	85.10	83.58
Nitrogen	0.22	0.40
Hydrogen	10.76	10.54

Experimental Set-up

A pyrex reactor with a quartz cover plate, high intensity light equipment (power supply, lamp housing and 1000 watt lamp), gas chromatograph, asymptotic calorimeter, three axis manipulator, type K thermocouple with digital thermometer, gas pump, safety bag, pressure gauge, and weighing machine capable of weighing to the nearest 0.0001 gram are the major equipment used during the course of this study. The rapid resid oil hydropyrolysis experiments are carried out in an experimental set-up as shown in Figure 1. The experimental set-up consists of a pyrex reactor, gas supply equipment, and a high intensity light assembly with a shuttering arrangement. Hydrogen and helium gases are supplied from high pressure (2500 psig) cylinders. Hydrogen gas is passed through several safety devices before entering the reactor as a precaution : (i) a flame arrestor is used to stop gas supply if a flashback occurs and to extinguish flame before it reaches the gas supply and (ii) a check valve to prevent any back-mixing. A pressure gauge is used to check for any leak in the system before each experiment is conducted as a safety measure. A teflon safety bag is used to prevent any pressure buildup during the experiment. Finally, a gas pump is used to purge the system with helium and hydrogen before each experiment.

92   Mathur, Salahuddin and Mohamed

<u>High intensity Light Assembly With Shuttering Arrangement.</u> The high intensity
light beam is produced by a xenon light equipment. The light source equipment
consists of two units: power supply and lamp housing. The power supply (LPS
1000) is designed to operate with high pressure xenon, mercury, or xenon-
mercury arc lamps in the range of 100 to 1000 watts and is capable of delivering
100 watt/cm^2 flux. The appropriate voltage for the lamp is automatically selected
and the power of the lamp is adjusted by varying the current. The instrument is
also equipped with a low operating circuit, in which case the lamp current can
be reduced to as low as 20 % of the normal value. Voltage and current are
displayed during each operation. The power supply is air cooled.

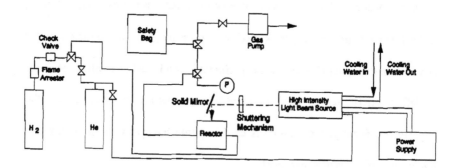

Figure 1: Schematic Diagram of the Experimental Set-up

The arc lamp housing is designed to accommodate selected xenon, mercury, or mercury-xenon arc lamp with power rating up to 1000 watts. The housing uses an f/4 elliptical reflector, which gives a high reflection efficiency. With the f/4 reflector, a horizontal beam of light is focused at a distance 407.1 mm from the front window. The front window is made of quartz so that the wavelengths of light greater than 250 nanometers (nm) can pass through. The lamp housing is water cooled and sealed, so that a venting system is not required. Also, it is designed for nitrogen or helium purging to prevent the production of ozone. The light source, a 1000 watt xenon arc lamp with a spectral emission in the range of 250 nm to 2500 nm, is assembled in the lamp housing. A xenon lamp is used in this study because its spectral emission is similar to that of sunlight. The lamp can be operated at various power levels and is capable of delivering up to 100 watt/cm^2 flux beam at 1000 watts power level.

A shuttering mechanism is designed and incorporated between the window and the reflective solid mirror to expose the resid oil sample for a few minutes at a time. A 45 degree reflective solid mirror is used to change the orientation of the high intensity light beam from horizontal to vertical. The reflective solid mirror is coated with magnesium fluoride to achieve maximum reflectance. The solid mirror is cooled by forced air to avoid the loss of the magnesium fluoride coating.

Reactor Design. The resid oil hydropyrolysis reactor is designed and constructed to have several characteristics that make this investigation unique. With this reactor, rapid heating rates of the resid oil and quenching of the resid oil volatiles and products in relatively cool surrounding gases and reactor wall are possible.

The hydropyrolysis reactions are conducted in a cylindrical pyrex reactor. The details of the hydropyrolysis reactor are shown in Figure 2. The reactor is 74 millimeters (mm) i.d., 80 mm o.d., and 122.1 mm high with a thicker wall at the top of the reactor to accommodate a 75 mm o-ring to provide gas seal. A quartz cover plate at the top of the reactor opening allows the high intensity light to pass through. The dimensions of the quartz plate are 101.6 mm diameter and 3.2 mm thick. The inlet and outlet tubing to the reactor are located at the bottom and top of the reactor, respectively. A gas sampling port is also located at the top of the reactor on the opposite side of the reactor outlet. A light beam from the high intensity light equipment moves horizontally. The beam is deflected at 90° by a reflective solid mirror and enters the reactor from the top through the quartz plate which serves as a lid as well as a window for the entering beam. A resid oil sample is placed in a low wide-form crucible. The dimensions of the crucible are 18 mm diameter at the top and 12 mm high. About 0.08 gram of resid oil is placed in the crucible which is placed on a perforated aluminum sheet of about 1 mm thick, which acts as a support. The resid oil sample is exposed to the radiant heat by positioning the sample at the

focal point of the beam. A gaseous product analysis shows that the gas sample

is rich in hydrogen indicating that the amount of hydrogen is adequate for the

hydropyrolysis of the resid oil sample.

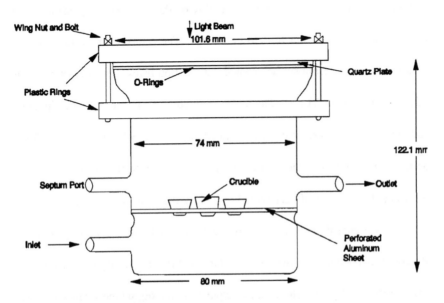

Figure 2: Structural Details of Hydropyrolysis Reactor

Hydrogen and helium atmospheres are used for the resid oil

hydropyrolysis and pyrolysis, respectively. Helium is used for several reasons.

First, the thermal conductivity of helium matches with that of hydrogen in the

temperature range of interest. Second, it is inert and is used for conducting

pyrolysis experiment as a comparison to those in hydrogen atmosphere. Finally,

it is used to purge the reactor several times to remove all the air before hydrogen is passed through. This is to avoid the mixing of air and hydrogen which can cause an explosion.

Heat Flux and Temperature Measurements. An asymptotic water cooled calorimeter is used to measure the heat flux of the high intensity beam at the focal point where the sample is to be located. The calorimeter sensor is made of copper with the front surface coated with carbon black. The diameter of the calorimeter is about 25 mm. The meter is mounted on a three axis manipulator and positioned to intercept the concentrated high intensity light beam at its focal point. The output from the calorimeter is sent to a voltmeter which is used to determine the amount of heat flux absorbed by the resid oil sample.

A digital thermocouple thermometer is used with a type K (chromel - alumel) thermocouple to measure the temperature of the high intensity light beam at the focal point. The voltage output from the digital thermometer is sent to a personal computer using a data acquisition program called LABTECH NOTEBOOK. The output is continuously recorded by the computer. Throughout this study, the gaseous and liquid samples are characterized using a gas chromatograph. Details of the analytical procedure are provided later.

Experimental Procedure

A resid oil sample weighing about 0.08 gram is placed in a porcelain crucible which is placed on a perforated aluminum sheet acting as a support in the reactor. The quartz plate lid is then clamped to the reactor. This assembly is then connected to the rest of the system.

The system is then purged with helium three or four times to expel any air and keep the contents of the reactor in a total helium atmosphere. The system is then purged with hydrogen to keep the reactor contents in hydrogen. This procedure is adopted to prevent any mixing of air with hydrogen which can lead to an explosion. The inlet and outlet to and from the reactor are then closed so that the resid oil rapid hydropyrolysis reaction can be carried out in a batch system. A pressure gauge is used to check for any leaks.

The light beam equipment is then turned on while the shutter is closed. The power is adjusted to a low level (less than 100 watts). The shutter is then opened to allow the beam to pass, enabling the sample to be positioned at the beam focal point. The shutter is closed again and the power is adjusted to a desired level. The shutter is then opened to expose the sample to rapid heating for a given period of time. The duration of light beam exposure is controlled by the shutter.

After the sample has been exposed, the reactor and its contents are allowed to cool down and a gas sample is taken through the septum port (before the quartz plate is removed). The gas sample is analyzed using the gas chromatograph. The quartz plate is then removed. Both the reactor and the quartz lid are then washed with analytical grade acetone. The acetone soluble product oil is then transferred into a sample bottle and is analyzed later. This is referred as low boiling point product. The reactor and the quartz lid are then washed with analytical grade tetrahydrofuran before another experimental run is conducted.

The liquid product sample (acetone soluble) is subjected to the ASTM D2887-89 standard test method for determining boiling point range distribution of various hydrocarbon fractions by a gas chromatograph. This procedure is also known as gas chromatographic simulated distillation technique for boiling point measurements. The gas is also analyzed using gas chromatograph with a different column.

After the simulated distillation of the acetone soluble product, acetone is evaporated from the sample using helium. The product oil is then weighed, labeled, and stored. Crucible with carbon residue and ash are also weighed, labeled, and stored.

Percent Resid Oil Conversion (Liquid Hydrocarbons + Gas)

The percent resid oil conversion (total yield) is calculated as :

$$\% \text{ Total conversion} = \frac{\text{initial weight of resid oil - weight of residue}}{\text{initial weight of resid oil}} \times 100$$

The percent resid oil converted to low boiling point liquid product (acetone soluble, b.p. < 425 °C) is calculated as:

$$\% \text{ Low b.p. liquid} = \frac{\text{weight of liquid (acetone soluble)}}{\text{weight of resid oil}} \times 100$$

The percent resid oil converted to high boiling point liquid product (tetrahydrofuran soluble, b.p. > 425 °C) is calculated as:

$$\% \text{ High b.p. liquid} = \frac{\text{weight of liquid (THF soluble)}}{\text{weight of resid oil}} \times 100$$

The percent of resid oil converted to gaseous product is calculated as:

$$\% \text{ Gases} = \% \text{ Total conversion} - \% \text{ Low b.p. liquid} - \% \text{ High b.p. liquid}$$

RESULTS AND DISCUSSIONS

The objective of this investigation was to study the rapid hydropyrolysis of Arabian Light atmospheric resid oil and vacuum resid oil for the production of light distillates. The results of this study have been divided into the effect of exposure time, temperature, and gaseous atmosphere. The heat flux used was in the range of 70 to about 97 watt / cm^2. The results from ASTM simulated distillation of the hydrogenated oil obtained at various experimental conditions are also presented.

Effect of Exposure Time

The effect of exposure time on the rapid hydropyrolysis of atmospheric and vacuum resid oils was studied in the range of one to six minutes at a constant temperature of 940 °C. This was the highest temperature which could be obtained in our equipment. A hydrogen atmosphere was maintained above the samples. The results are shown in Figure 3. As can be seen, the total conversion for atmospheric resid oil begins at 40.5% for one minute, reaching a value of 84.4% at three minutes and then remaining almost constant. Similar results are seen for low boiling point products (acetone soluble) which reach a value of 56.4% at three minutes exposure time with no appreciable increase with further increase in exposure time.

In another set of experiments, vacuum resid oil was subjected to rapid hydropyrolysis under similar experimental conditions. The results are shown in Figure 4. The total conversion is 78.4% with a yield of low boiling point products of 47.5% when the sample is subjected to three minutes of hydropyrolysis. The conversions of the vacuum resid oil are less, as expected, due to the asphaltic nature of the resid oil with no volatile matter below 560 °C.

However, these conversions at atmospheric pressure with no catalyst usage are

very attractive.

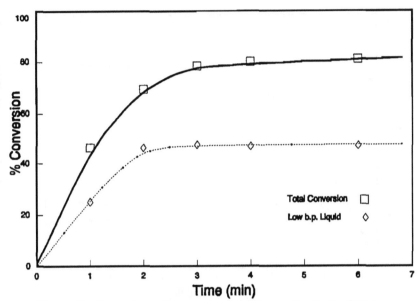

Figure 3 : Percentage Conversions Of Atmospheric Residual
Oil versus Time of Exposure at 940C

Production of Low Boiling Point Products (Acetone Soluble). In the

rapid hydropyrolysis, the sudden increase in temperature causes primary thermal

cracking reactions. Because of the low temperature of the resid oil surroundings,

the molecules do not go through secondary cracking but instead react with

hydrogen to produce low boiling point molecules. However, if the exposure time

is increased, the surrounding environment gets heated and causes secondary

cracking. A product distribution in Figure 5 shows that for atmospheric resid

oil, the low boiling point fraction (less than 300 °C) is about 14.3% at one

Mathur, Salahuddin and Mohamed

minute exposure time, increases to 17.3% at three minutes, and then reduces to

7.5% at six minutes. The low amount of low boiling point fraction at one minute

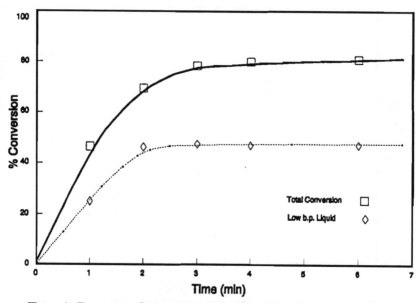

Figure 4 : Percentage Conversions of Vacuum Residual Oil
versus Time of Exposure at 940C

is because of the inadequate time available for thermal cracking. Similar results

are observed for vacuum resid oil providing 13.4% of low boiling point fraction

in the kerosine range (less than 300 °C) at three minutes of exposure time

(Figure 6).

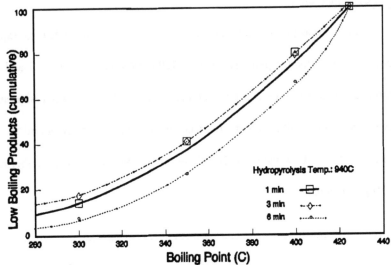

Figure 5 : Low Boiling Products Distribution (cumulative) of
Atm. Residual Oil at Residence Time of 1,3 and 6 minutes

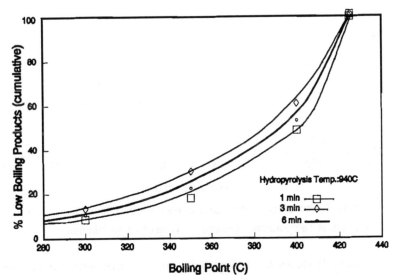

Figure 6 : Low Boiling Products Distribution (cumulative) of
Vacuum Residual Oil at Residence Time of 1,3 and 6 minutes

Effect of Temperature

Both atmospheric and vacuum resid oils were subjected to rapid
hydropyrolysis in the temperature range of 700 to 940 °C (see Figures 7 and 8)
keeping the exposure time constant at three minutes. The total conversion and
conversion to low boiling point products for both resid oils are found to be
highest at 940 °C. Low values at 700°C are due to inadequate primary thermal
cracking reactions.

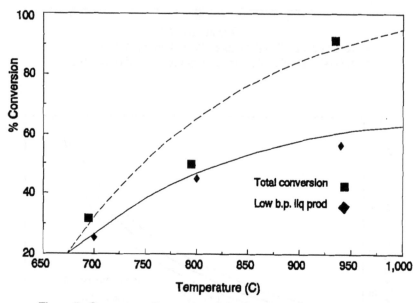

Figure 7 : Percentage Conversion of Atmospheric Residual
Oil versus Temperature at Exposure Time of Three Minutes

Table II shows the boiling point product distribution for the rapid
hydropyrolysis of the two resid oils at 700 and 940 °C. The atmospheric resid

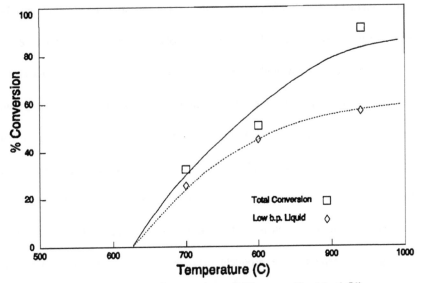

Figure 8 : Percentage Conversions Of Vacuum Residual Oil
versus Temperature at Exposure Time of Three Minutes

oil produces low boiling fractions (less than 300°C and 300 - 350 °C) in higher

amounts than obtained from vacuum resid oil. However, a production of 13.4%

in the kerosine boiling range (less than 300 °C) from vacuum resid without the

use of high pressure equipment or a catalyst is very attractive.

Effect of Helium and Hydrogen Atmosphere

Experiments were conducted on atmospheric and vacuum resid oils at

940 °C for three minutes exposure time in helium and hydrogen atmosphere.

Results from these experiments are shown in Table III. The total conversion for

hydrogenated to low boiling point products. On the other hand, when similar experiments were conducted in helium the total conversion was about 88.8% with conversion to low boiling point products of only 44.7%. Similarly for vacuum resid oil, percent total conversion and conversion to low boiling point products were found to be 78.4 % and 47.5%, respectively in hydrogen. The conversions were reduced to 77.7% and 32.0% in helium atmosphere. Almost the same percent total conversion is observed for both resid oils in hydrogen and helium atmospheres which is considered due to equal primary cracking. In a hydrogen atmosphere, the products from the primary thermal cracking are stabilized by hydrogen yielding higher amounts of low boiling point products. However, in a helium atmosphere the primary products repolymerize with one another to produce higher boiling point products. The low boiling point product distributions for the two resid oils in hydrogen and helium atmospheres are presented in Table IV.

Table II

Low boiling product distributions of hydropyrolized resid oils for three minutes at different temperatures.

Boiling point	Atmospheric residual oil		Vacuum residual oil	
	700°C	940°C	700°C	940°C
< 300°C	16.5	17.3	13.0	13.4
300-350°C	30.6	23.7	18.1	16.6
350-400°C	38.4	38.1	29.3	31.0
400-425°C	14.5	20.9	39.6	39.0

Table III

Percentage conversion of resid oils in helium and hydrogen atmosphere at temperature of 940°C and a residence time of three minutes.

	Percentage conversion of			
	Atm. residual oil		Vacuum residual oil	
	overall	low bp. liq. prod.	overall	low bp. liq. prod.
Helium	88.8	44.7	77.7	32.0
Hydrogen	84.4	56.4	78.4	47.5

Table IV

Low boiling product distributions of resid oils pyrolyzed in helium and hydrogen atmospheres at 940°C and three minutes.

| | Atm. residual oil | | Vacuum residual oil | |
	hydropyrolysis	pyrolysis	hydropyrolysis	pyrolysis
Boiling Point				
<300°C	17.3	7.7	13.4	8.9
300–350°C	23.7	20.4	16.6	16.8
350–400°C	38.1	40.8	31.0	51.9
400–425°C	20.9	31.1	39.0	22.4

CONCLUSIONS

The following conclusions can be made:

1) The experimental set-up successfully conducted rapid hydropyrolysis of atmospheric and vacuum resid oils.

2) For the atmospheric resid oil, the percentage total conversion of 84.4% and percent conversion to low boiling point products of 56.4% are obtained at 940 °C and three minutes exposure time in hydrogen atmosphere.

3) For the vacuum resid oil, the percentage total conversion of 78.4% and percent conversion to low boiling point products of 47.5% are obtained at 940 °C and three minutes exposure time in hydrogen atmosphere.

4) The maximum amount of low boiling point liquid products in the kerosine boiling range (b.p. less than 300 °C) is 17.3% for atmospheric resid oil

and is obtained at 940 °C and three minutes exposure time in hydrogen atmosphere.

5) The maximum amount of low boiling point liquid products in the kerosine boiling range (b.p. less than 300 °C) is 13.4% for vacuum resid oil and is obtained at 940 °C and three minutes exposure time in hydrogen atmosphere.

6) The rapid hydropyrolysis of atmospheric and vacuum resid oils can result in high percentage total conversion and conversion to low boiling point products without the use of high pressure and catalyst.

Acknowledgement

Authors thank Mobil Corporation, Princeton, NJ for providing residual oil samples.

LITERATURE CITED

1. Gary, J.H., and Handwerk, G.E., "Petroleum Refining, Technology and Economics," Marcel Dekker, New York, 1975.

2. Speight, J.G., "The Desulfurization of Heavy Oils and Residua," Marcel Dekker, New York, 1981.

3. Lopez, J., McKinney, J.D., and Pasek, E.A., "Heavy Oil Hydroprocessing," U.S. Patent No. 4557821, Dec. 10, 1985.

4. Mathur, V.K., and Venkataramanan, V., Preprint A.C.S. Division of Fuel Chemistry, Vol. 27, No. 2, 1982, pp. 1.

5. Mathur, V.K., Fakoukakis, E.P., and Ruether, J.A., Fuel, Vol. 63, 1984, pp. 1700.

6. Mathur, V.K., and Reddy Karri, S.B., Fuel, Vol.65, 1986, pp. 790.

7. Mohamed, A.R., and Mathur, V.K., Fuel, Vol. 70, 1991, pp. 983.

8. Beattie, W.H., and Sullivan, J.A., Proceedings of the 15th Intersociety Energy Conversion Engineering Conference, 1980, pp. 637.

9. Pyatenko, A.T., Bukhman, S.V., Lebedinskii, V.S., Nasarov, V.M., and Tolmachev, I.Y., Fuel, Vol. 71, 1992, pp. 701.

10. Beattie, W.H., Berjoan, R., and Coutures, J.P., Solar Energy, Vol. 31, No. 2, 1983, pp. 137.

11. Personal Communication, Scholl, K., "Coal Liquefaction Using Solar Thermal Energy,"National Renewable Energy Laboratory, Golden, CO, 1992.

12. Ballantyne, A., Chou, H., Neoh, K. Orazco, N., and Stickler, D., AIChE National Meeting, St. Louis, MO., March 1984.

13. Fallon, P.T., and Steinberg, M., Proceedings of the 16th Intersociety Energy Conversion Engineering Conference, Vol. 2, 1981, pp. 1106.

14. Sugawara, T., Sugawara, K., Sato, S., Chambers, A.K., Kovacik, G., and Ungarian, D., Fuel, Vol. 69, 1990, pp. 1177.

15. Niksa, S., Heyd, L.E., Russel, W.B., and Saville, D.A., Twentieth Symposium (International) on Combustion, The Combustion Institute, Pittsburgh, 1984, pp. 1445.

16. Mohamed, A.R., Ph.D. Thesis, University of New Hampshire, 1993.

6 Residue Upgrading by Hydrovisbreaking and Hydrotreating

Stuart S. Shih

Mobil Research and Development Corporation

Paulsboro, N.J. 08066-04800

ABSTRACT

The effect of hydrovisbreaking (HVB) of Arabian Light vaccum residue on the performance of resid demetalation and desulfurization catalysts was studied. The HVB/HDT combination results in higher conversion with reduced viscosity compared to catalytic hydrotreating alone. This approach can be attractive for the production of low-sulfur heavy fuel oils with reduced cutter stock requirements or when the goal is conversion in combination with hydrodemetalation and desulfurization. However, hydrovisbreaking does not improve performance of the downstream catalysts in terms of desulfurization and demetalation. In addition, hydrovisbreaking at a too high conversion can result in sedimentation which can plug downstream reactors or cause rapid pressure-drop increase. Therefore, the HVB/HDT combination would not be suitable for a fixed-bed residue hydrotreating process where high conversion is not a primary goal.

111

INTRODUCTION

Recent literature reports that cascade hydrovisbreaking/hydrotreating (HVB/HDT) can improve overall residue conversion and reduce residue viscosity without excess sediment formation as long as the hydrovisbreaking and overall conversion is below a critical level[1-5]. This approach has been reported to maintain high conversion without subjecting the hydrotreating catalysts to high temperature at start of the cycle conditions [1,2]. The viscosity reduction reduces the cutter stock requirement for producing heavy fuel oils. Hydrovisbreaking can be accomplished either in a feed pre-heater or in an empty reactor (high-pressure soaker drum). However, the effect of hydrovisbreaking on subsequent catalytic performance and operating feasibility (e.g., pressure drop or reactor plugging) is not clear. This work was done to determine the effect of hydrovisbreaking on subsequent hydrotreating catalyst performance.

EXPERIMENTAL

Desulfurization and demetalation catalysts were evaluated separately with and without hydrovisbreaking. The experiments were conducted in a dual reactor pilot unit. The first reactor was filled with 14/60 mesh quartz to simulate hydrovisbreaking; and the second reactor was filled with a hydrotreating catalyst. For hydrotreating only, the first reactor was kept at 300°F so that

residue could flow through the reactor without any

hydrovisbreaking. In the HVB/HDT case, the first reactor

temperature was raised to 760°F. The residence time of oil in the

first reactor was about 1.0 hour or equivalent to an ERT (equivalent

reaction time) of 1080 seconds for hydrovisbreaking at 1900 psig

H_2. ERT, known also as the soaking factor, was frequently used to

express the severity of visbreaking in terms of seconds at 800°F [6].

The visbreaking severity includes the reaction time and

temperature. For the visbreaking at temperatures other than the

reference temperature of 800°F, ERT is calculated by using an

activation energy of 50 kcal/mole. The hydrotreating catalysts used

in this study were a small-pore $NiMo/Al_2O_3$ desulfurization

catalyst and a large-pore $NiMo/Al_2O_3$ demetalation catalyst (Table

1). After presulfiding and stabilizing with the feed, Arabian Light

vacuum residue at 600-700°F for one week, the catalyst activities

were measured at 720°F, 740°F, and 760°F (0.35 LHSV, 1900 psig

H_2). The catalyst activity checks were repeated after the first

reactor was raised to 760°F to obtain data for the HVB/HDT

combinations cases. All experiments were conducted with Arabian

Light vacuum residue as the feedstock (Table 2).

TABLE 1   Fresh Catalyst Properties

Catalyst	Small-Pore	Large-Pore
Nickel, wt %	2.5	1.5
Molybdenum, wt %	10.0	3.6
Particle Density, g/cc	1.30	1.1
Pore Volume, cc/g	0.491	0.66
Surface Area, m^2/g	162	145
Avg Pore Dia., Angstroms	118	181
Pore Size Distribution, cc/g (H_g porosimetry)		
<50 Angstroms	0.02	0.00
50-100 Angstroms	0.33	0.09
100-300 Angstroms	0.15	0.58
>300 Angstroms	0.01	0.02
Total	0.51	0.69

TABLE 2   Feed: Arabian Light Vacuum Residue

Gravity, API	7.6
Hydrogen, wt %	10.84
Sulfur, wt %	4.2
Nitrogen, wt %	0.25
CCR, wt %	18.97
Asphaltenes, wt %	13.92
Trace Metals, ppmw Nickel	19
Vanadium	75
Iron	13
Sodium	22
Viscosity KV @ 212°F, cs	730
Composition, wt % 650°F⁻	3
650-1000°F	10
1000°F⁺	87

RESULTS AND DISCUSSION

Viscosity Reduction

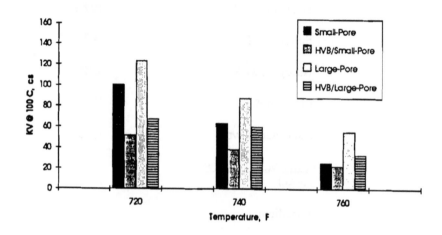

Figure 1. Effects of Hydrovisbreaking on Product Viscosities

Hydrovisbreaking alone reduced viscosity of the Arabian Light vacuum residue from 730 cs to 184.4 cs (KV @ 100°C) at conditions conducted in this study (Table 3).

TABLE 3   Hydrovisbreaking Performance

Desulfurization, %		9.5
Denitrogenation, %		0
Demetalation, %		2.0
CCR Reduction, %		6.1
Asphaltene Conversion, %		10.6
$1000^\circ F^+$ Conversion, %		4.5
Viscosity	Feed	Product
KV @ $130^\circ F$, cs	-	3733
KV @ $212^\circ F$, cs	730	184.4

Conditions:
1.0 hour, $760^\circ F$, 1900 psig.

As expected, no significant desulfurization and demetalation were
observed for the hydrovisbreaking alone case. Combined with the
hydrotreating step, viscosities of the upgraded residua from the HVB/HDT
combination were 50-60% of those from the HDT alone case (Figure 1). The
small-pore catalyst, which was more active for hydrogenation (Figure 2) due to
its higher surface area and higher NiMo loading, provided better overall
viscosity reduction than the large-pore catalyst (Figure 1). For the HVB/HDT
combination, the incremental conversion (4.5 % $1000^\circ F^+$ conversion) by

hydrovisbreaking is responsible for the overall improvement in the viscosity

reduction (Figure 3). As shown in Figure 4, viscosity of the upgraded residua

is a function of overall 1000°F⁺ conversion only. Catalyst type has little

effect on the viscosity/conversion selectivity although the HVB/HDT

combination for the small-pore catalyst results in higher viscosity reduction. It

is expected because viscosity is strongly dependent on size of residue

molecules which decreases with increasing 1000°F⁺ conversion.

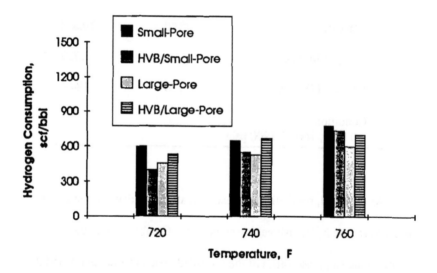

Figure 2 Effects of Hydrovisbreaking on Hydrogen Consumption

Sediment Formation

No measurable sediments (toluene insolubles) were observed in this study because of the relatively low conversions by both the HVB and HDT steps. Attempts were made to increase HVB severity by raising the first reactor temperature from $760^{o}F$ to $800^{o}F$. However, the pressure drop increased rapidly and the reactor was completely plugged after one day at $800^{o}F$. Of course, such plugging would not be a problem if an open tube (e.g., furnace) or an empty reactor (soaker drum) was employed in the HVB step. Nevertheless, this experiment did demonstrate the sensitivity of fixed-bed reactors to sediments (coke) in the oil. Even the HVB step was carried out in the furnace, the HVB severity and sediments had to be carefully controlled to avoid plugging in the downstream fixed-bed unit. Furthermore, the benefits of viscosity reduction and residue conversion are not so important if the hydrotreated residue will be further upgraded in an FCC unit to gasoline and distillates. Therefore, the HVB/HDT combination should not be recommended for the fixed-bed processes for such applications. However, it is known that addition of H-donor solvents, and dispersed catalysts to vacuum residua could suppress the coke formation and might be able to maintain high conversions without reactor plugging [7,8,9].

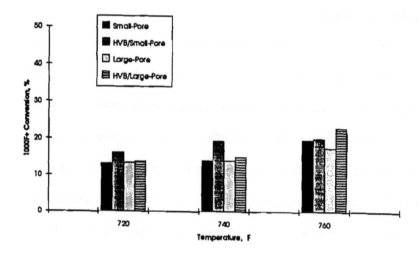

Figure 3   Effects of Hydrovisbreaking on Residue Conversion

HDT Performance

The HVB pretreatment step did not improve subsequent catalytic

performance in terms of desulfurization (Figure 5), demetalation (Figure 6),

CCR reduction (Figure 7), and asphaltene conversion (Figure 8) for either

small-pore or large-pore catalysts. Most data even suggest that the

hydrovisbroken residue in the HVB/HDT combination is more refractory than

the virgin residue in the HDT only case. This is not surprising since

hydrovisbreaking, similar to thermal visbreaking, can produce and concentrate

refractory asphaltenes concentrated with metals and polar functional groups

[3,10]. However, hydrovisbreaking appears to improve the hydrogenation

activity of the large-pore catalyst as shown in Figure 2. The reasons for this improvement are not clear. It is possible that the hydrovisbreaking reduces size of the residue molecules and lowers the diffusion resistance. However, this improvement was not observed for the demetalation which is a diffusion-controlled reaction.

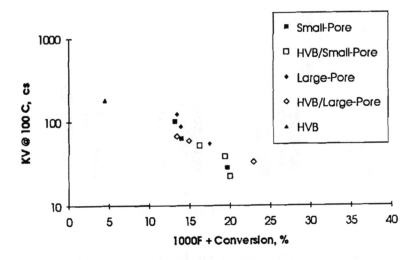

Figure 4   Plotting of Viscosity vs Residue Conversion

At controlled HVB conditions, the HVB/HDT combination can achieve higher conversion without excess sediments generated by hydrovisbreaking [3]. The amount of sediments formed generally can be correlated with the asphaltene content of upgraded products. Consequently, conditions that achieve less asphaltenes conversion tend to produce more sediments in the products. As shown in Figure 9, the HVB/HDT combination allows higher conversion at constant asphaltene content (or constant asphaltene conversion). This improvement was probably due to the incremental $1000^{\circ}F^{+}$ conversion (4.5%) generated in the hydrovisbreaking step.

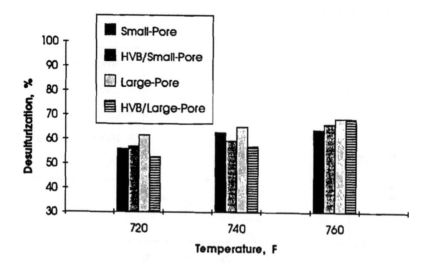

Figure 5   Effects of Hydrovisbreaking on Desulfurization Activity

CONCLUSIONS

The HVB/HDT combination can improve overall residue conversion and reduce residue viscosity without excess sediments. This approach can be attractive for the production of low-sulfur heavy fuels with reduced cutter stock requirements. However, hydrovisbreaking does not improve subsequent catalytic performance of the hydrotreating catalysts. In addition, uncontrolled hydrovisbreaking can produce sediments that can lead to reactor plugging or rapid pressure-drop -increase.

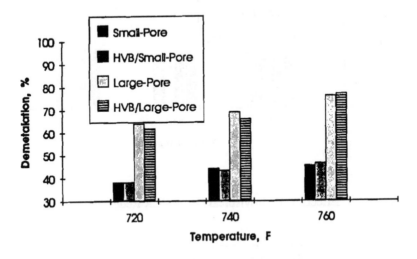

Figure 6 Effects of Hydrovisbreaking on Demetalation Activity

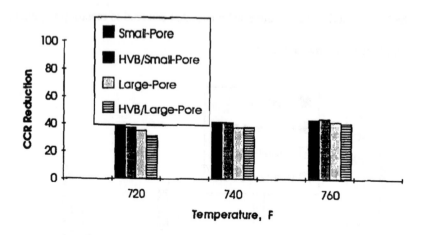

Figure 7   Effects of Hydrovisbreaking on CCR Reduction Activity

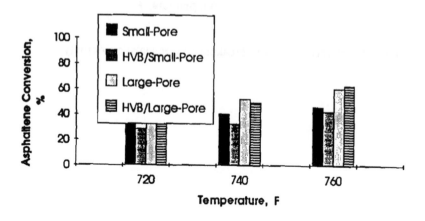

Figure 8   Effects of Hydrovisbreaking on Asphaltene Conversion Activity

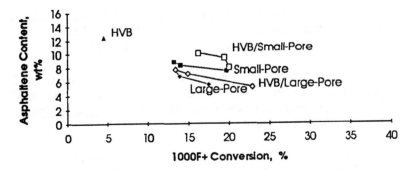

Figure 9 Plotting of Asphaltene Content versus Residue Conversion

REFERENCES

1.   L. Mariette, A. Billon, and T. des Courieres, "HVYAHL T. Process for High Conversion of Resids,", paper presented at the Japan Petroleum Institute Petroleum Refining Conference, Tokyo, October 19-21, 1988.

2.   J. P. Peries, C. Raimbault, J. des Courieres, and L. Gouzien, "TERVAHL Process at the Border Between Thermal and Catalytic Resid Upgrading Process," paper presented at the 1990 NPRA Annual Meeting, AM-90-26, San Antonio, Texas, March 25-27, 1990.

3.   T. Takatuka, Y. Wada, Y. Fukui, S. Komatsu, and S. Shimizu, "VisABC Process: Practical Candidate to Maximize Residue Conversion," paper presented at the 1988 NPRA Annual Meeting, AM-88-63, San Antonio, Texas, March 20-22, 1988.

4.   T. Takatsuka, Y. Wada, Y. Shiroto, Y. Fukui, and S. Komatsu, "Latest Development in the ABC Process," paper presented at the Japan Petroleum Institute Petroleum Refining Conference, Tokyo, October 19-21, 1988.

5.   S. S. Shih, "Summary of Hydrodemetalation Technology," paper presented at AIChE Annual Meeting, Chicago, November 11-16, 1990.

6.   T. Y. Yan, "Coke Formation in Visbreaking Process," paper presented at ACS Meeting, Denver, April 5-10, 1987.

7.   R. H. Heck, L. A. Rankel, and F. T. Quiseppi, "Conversion of Petroleum
        Resid from Maya Crude: Effects of H-donor, Hydrogen Pressure and
        Catalyst," Fuel Processing Technology, vol. 30, pp. 69-81, 1992.

8.   D. Decroocq, B. Fixari, M. Bigois, T. Des Courieres, J. Rossarie, and L.
        Lena, "Improving Process of Thermal Cracking of Petroleum Residua by
        Optimizing Conversion of Resins and Asphaltenes," Revue de I'Institut
        Francasis du Petrole, vol. 47, #1, pp. 103-131, 1992.

9.   M. J. Dabkowski, S. S. Shih, and K. R. Albinson, "Upgradin of
        Petroleum Residue with Dispersed Catalysts," in AIChE Symposium
        Series 282 volume 87, 1991: Tar Sand and Oil Upgrading Technology,
        edited by S. S. Shih and M. C. Oballa.

10.   D. D. Whitehurst, T. O. Mitchell, and M. Farcasiu, Coal Liquefaction:
        the Chemistry and Technology of Thermal Processes, Academic Press,
        Inc., New York, 1980.

7 Mild Hydrocracking of Heavy Oils with Modified Alumina Based Catalysts

E. P. Dai and C. N. Campbell

Texaco Research and Development
P. O. Box 1608
Port Arthur, Texas 77642

INTRODUCTION

The decreasing demand for heavy fuel oils requires that refiners find ways for converting heavy hydrocarbon feedstocks to higher value mid-distillate products. To increase mid-distillate production, the refiner can choose from several processing options such as hydrocracking, fluid catalytic cracking, and coking. All of these options, however, require heavy capital investments. Because of these high investment costs, refiners are continually searching for conversion processes which may be utilized in existing units. One such process is mild hydrocracking (MHC).

The major advantage of MHC is that it can be carried out within the operating constraints of existing hydrotreating units. The typical operating conditions of VGO hydrotreaters are: Temperature: 720-780F (382-415C), Hydrogen pressure: 600-1200 psig (4137-8274 kPa), H_2/Oil ratio: 1000-2000 SCF/BBL (0.2-0.4 m^3/l), and Space Velocity: 0.4-1.5 vol/vol/hr. In contrast, true hydrocracking units are operated at the conditions: Temperature: 780-900F (415-482C), Hydrogen pressure: 1800-3000 psig (12,411 - 20,685 kPa), H_2/Oil ratio: 1400-6000 SCF/BBL (0.28-1.2 m^3/l), and Space Velocity: 0.3-1.5. By increasing the operating severity in an existing unit a heavy gas oil hydrotreater can be used for the MHC process.

Using a conventional hydrotreating catalyst, the MHC process typically converts about 10 to 30 vol% of a hydrocarbon feedstock boiling above 670 F (670 F+, 354 C+) to middle distillates boiling at or below 670 F (670 F-, 354 C-). For a residuum feedstock, it usually gives less than 10 vol% conversion of the 670 F+ (354 C+) fraction. With the known alumina-based hydrotreating catalysts the conversion of resid components boiling above 1000 F (1000 F+, 538 C+) into products boiling at or below 1000 F (1000 F-, 538 C-) is achieved primarily by thermal cracking reactions. A difficulty which arises in resid

127

hydroprocessing units employing the currently known catalysts is the formation of insoluble carbonaceous substances (also called sediment) when the conversion is high (above 50 vol%). The higher the conversion level for a given feedstock, the greater the amount of sediment formed.

The MHC catalysts in the patent literature (Ref. 1-5) generally contain acidic cracking components such as hydrogen Y and ammonium exchanged Y zeolites as well as silica-alumina. In addition, the catalyst may also contain a halogen element, such as fluorine, for enhancing the cracking activity. NiMoP catalysts on alumina supports and NiMo catalysts on boron-promoted alumina supports have also been reported to be effective for the mild hydrocracking of heavy oils.

The general objective of this work is to identify an MHC catalyst which gives a higher conversion level for heavy hydrocarbon feedstocks, especially that fraction of the feedstock that boils above 1000 F (538 C), while maintaining the same amount of sediment production.

EXPERIMENTAL

A. Mild Hydrocracking Catalyst Evaluation

The Berty reactor, a type of continuous stirred tank reactor (CSTR), was used to determine mild hydrocracking activities of the candidate catalysts in a diffusion controlled regime at a low rate of deactivation. After being loaded in the reactor, the catalyst was presulfided and then the reaction was carried out at a single space velocity for 38 hours. Sample cuts were taken every 4 hours and tested for boiling point distribution, and nickel, vanadium, sulfur, and sediment content. Using these data, conversions for the 650 F+ (343 C+) and 1000 F+ (538 C+) fractions were determined. The feedstock properties and the operating conditions for the reactor evaluations are listed in Table I.

TABLE I

BERTY REACTOR OPERATING CONDITIONS

I. PRESULFIDING

Temperature	750 - 800 F (399 - 427 C)
Pressure	40 psig (276 kPa)
Gas Mixture	10 vol% H2S - 90 vol% H2
Gas Flow	500 sccm
Duration	2 hr 45 min

II. FEEDSTOCK   60 vol% Desulfurized VGO
                                 40 vol% Ar M/H Vac. Resid

Boiling Point	IBP	444 F (229 C)
Distribution	FBP	1371 F (744 C)
	650 F+(343 C+)	89.2 vol%
	900 F+(482 C+)	45.6 vol%
	1000 F+(538 C+)	33.5 vol%
Sulfur wt%	2.2	
Ni Content, ppm	20	
V Content, ppm	54	

III. REACTION CONDITIONS

Temperature	805 F (429 C+)
Pressure	1000 psig (6895 kPa)
H2 Feed Rate	300 sccm
Liquid Feed Rate	82.5 cc/hr
Liquid Holdup	125 cc
Catalyst Charge	36.9 grams

The mild hydrocracking activity was determined by comparing the percentages of products in the 650 F- (343 C-) fraction and 1000 F- (538 C-) fraction when various catalysts were evaluated under constant mild hydrocracking conditions with the same feedstock. The conversions of the 650 F+ (343 C+) and 1000 F+ (538 C+) fractions were calculated using the equation:

$$Conversion = \frac{Y(F) - Y(P)}{Y(F)} \; X \; 100$$

where

$Y(F)$ = volume percentage of the 650 F+ or 1000 F+
fraction in the feedstock

$Y(P)$ = volume percentage of the 650 F+ or 1000 F+ fraction
in the product.

B. Catalyst Preparations

The catalysts were prepared from commercially available porous supports composed of alumina, boria-alumina, magnesia-alumina, silica-alumina, titania-alumina, or Y-zeolite-alumina. Tables IV, V and VII list the properties of these supports including Total Surface Area (TSA) in square meters per gram of the support and Total Pore Volume (TPV) in cubic centimeters per gram. All of the supports were obtained from American Cyanamid Inc. and were extrudates with diameters of 0.035-0.041 inch (0.089-0.104 cm).

Each support was impregnated with the requisite amounts of Group VIB and VIII metal oxides and phosphorus oxide to yield a finished catalyst containing a Group VIII metal oxide in amount of 3-3.5 wt%, a group VIB metal oxide in amount of 14.5-16.5 wt% and phosphorus oxide in amount of 0-1.5 wt%. Stabilizers such as phosphoric acid, hydrogen peroxide and citric acid monohydrate, were employed to vary the surface distributions of active metals. The impregnated support was then oven-dried and calcined at 1000-1150 F (538-621 C) for 20 minutes to 2 hours in flowing air. The porosity data and chemical compositions of the finished catalysts are shown in Tables III-VIII.

RESULTS AND DISCUSSION

A. Catalyst Properties

The key catalyst properties are presented in Tables II - VIII. Catalyst A and Catalyst B, shown in Table II, represent the commercial NiMo catalysts on alumina supports with bimodal pore size distributions. HDS-2443 is a commercial NiMo catalyst on an alumina support with a monomodal pore size distribution.

TABLE II

ALUMINA BASED CATALYSTS AS CONTROL EXAMPLES

Catalyst	Catalyst A	HDS-2443B	Catalyst B
Impreg. Sol'n	Ni-Mo	Ni-Mo	Ni-Mo
MoO_3 wt%	11.5-14.5	14.5-15.5	9.8
NiO wt%	3.2-4.0	3.0-3.5	2.6
Pore Volume Distribution by Hg Porosimetry			
Total PV, cc/g	0.74	0.64	0.86
PM at (dv/dD)max Å	50	126	98
PM (BET), Å	46	105	112
Surf. Area, m^2/g	314	194	172
HDS-MAT, $C_{0.5g}$, %	73	88	79
Metals Distribution by XPS Analysis			
$(Mo/Al)_{int}$	0.09	0.012	0.09
$(Ni/Al)_{int}$	0.012	0.016	0.013
Mo Gradient	1.2	3.1	0.81
Ni Gradient	1.6	1.0	1.0

SN-6001 and SN-6010, shown in Table III, are NiMoP and NiMo alumina based catalysts, respectively. These catalysts have a pore mode of 110 Å and about 0.02 cc/g of macroporosity, which we define as the volume of pores with diameters greater than 250 Å (denoted as PV > 250 Å, cc/g). These catalysts also have a narrow pore size distribution with at least 75% of pore volume in pores with diameter between 100-160 Å. SN-6264 and SN-6224 catalysts have pore mode equivalent to SN-6001, but they have medium macroporosities between 0.06-0.10 cc/g.

TABLE III

NiMo CATALYSTS ON ALUMINA SUPPORTS

Catalyst	SN-6001	SN-6010	SN-6264	SN-6224
Impreg. Solution	Ni-Mo-P	Ni-Mo	Ni-Mo-P	Ni-Mo-P
P wt%	0.72	0.02	0.74	0.74
MoO_3 wt%	15.2	13.6	14.8	15.0
NiO wt%	2.9	2.9	3.1	3.1
TPV, cc/g	0.54	0.59	0.63	0.69
PM at (dv/dD)max Å	121	121	121	121
PM (BET), Å	110	110	110	110
Surf. Area, m^2/g	180	179	185	182
HDS-MAT, $C_{0.5g}$, %	86	78	98	92
$(Mo/Al)_{int}$	0.12	0.12	0.13	0.14
$(Ni/Al)_{int}$	0.014	0.016	0.012	0.011
Mo Gradient	1.1	3.0	1.3	1.5
Ni Gradient	1.0	1.0	1.5	1.2

The properties of NiMoP catalysts on the zeolite/alumina supports are given in Table IV. SN-6571 is a NiMoP catalyst supported on a carrier that consists of 80 wt% of a precipitated alumina and 19 wt% of an ultrastable Y zeolite with a unit cell size about 24.41 Å. SN-6571 has a pore mode of 98 Å and a macroporosity of 0.13 cc/g and pore volume of 0.25 cc/g in pores with diameters in the range of 100-160 Å.

TABLE IV

NiMoP CATALYSTS ON ZEOLITE/ALUMINA SUPPORTS

Catalyst	SN-6571X	SN-6571	SN-6572X	SN-6572
Impreg. Solution	Support	Ni-Mo-P	Support	Ni-Mo-P
P wt%		0.67		0.66
MoO_3 wt%		15.1		15.0
NiO wt%		2.9		2.9
TPV, cc/g	0.78	0.59	0.77	0.61
PM at (dv/dD)max Å	86	105	86	110
PM (BET), Å	99	98	99	94
Surf. Area, m^2/g	318	216	306	242
HDS-MAT, $C_{0.5g}$, %		79		87
$(Mo/Al)_{int}$		0.14		0.14
$(Ni/Al)_{int}$		0.016		0.016
Mo Gradient		1.2		1.3
Ni Gradient		0.9		1.3
Zeol. Content, wt%	19	15	16	17
UCS, Å	24.48	24.41	24.29	24.29
SiO_2/Al_2O_3	10	12	32	32

SN-6572 was prepared using a support comprised of 80 wt% precipitated alumina and 16 wt% of a dealuminated Y zeolite having a unit cell size of 24.29 Å and a silica-alumina molar ratio of about 32. The dealuminated Y zeolite has a secondary pore mode of about 85 Å plus, a higher pore volume in pores with diameter greater than 50 Å, and a lower acid site density relative to the ultrastable Y zeolite.   The pore modes of SN-6572 and SN-6571 are about equivalent, but the pore volume in PV, 100-160 Å is greater for SN-6572 (52% vs. 42% of TPV). SN-6572 also has a higher MAT activity than SN-6571.

SN-6412 and SN-6650 (Table V) are boria-silica-alumina based NiMo catalysts. SN-6412 has a pore mode of 116 Å, a macroporosity of 0.09 cc/g, a pore volume in PV 100-160 Å of 0.30 cc/g (57% of TPV) and a TPV of 0.53 cc/g. SN-6412 also has a low total surface area and the high molybdenum gradient (poor Mo dispersion). As a result, SN-6412 has a very low HDS-MAT activity. In contrast to SN-6412, SN-6650 has much improved HDS-MAT activity.

TABLE V

NiMo CATALYSTS ON BORIA-ALUMINA SUPPORTS

Catalyst	SN-6412X	SN-6412	SN-6650X	SN-6650
Sample	Support	Catalyst	Support	Catalyst
Impreg. Sol'n		H_2O_2		H_2O_2
B_2O_3 wt%		8.1		8.7
MoO_3 wt%		12.7		11.2
NiO wt%		2.8		3.1
Pore Volume Distribution by Hg Porosimetry				
TPV, cc/g	0.69	0.53	0.81	0.71
PV > 250A cc/g	0.12	0.09	0.09	0.07
PV > 160A cc/g	0.25	0.20	0.23	0.19
PV < 160A cc/g	0.44	0.34	0.59	0.53
PV < 100A cc/g	0.06	0.04	0.08	0.06
PV 100-160A cc/g	0.38	0.30	0.51	0.47
as % of TPV	55	57	63	66
MPD (Vol), A	139	139	137	135
TSA (N_2), m²/g	147	114	189	174
Metals Distribution by XPS Analysis				
Mo Gradient		17.8		1.9
Ni Gradient		4.4		1.7

Shown in Table VI are two other catalysts SN-6262 and SN-6411. SN-6262 is a NiMoP catalyst on a support made of 8 wt% silica and 92 wt% alumina, and SN-6411 is a NiMo catalyst on a support made of 10 wt% titania and 90 wt% alumina. As shown in Table VI, SN-6262 has a pore mode of 76 Å, a medium macroporosity of 0.14 cc/g, and about 15% of TPV in the PV, 100-160 Å as well as a TPV of 0.75 cc/g. SN-6411 has a pore mode about 130 Å, a very low macroporosity of 0.02 cc/g and about 75% of pore volume in PV, 100-160 Å.

TABLE VI

NiMo CATALYSTS ON SILICA AND TITANIA BASED SUPPORTS

Catalyst	SN-6262	SN-6411
Impreg. Sol'n	Ni-Mo-P	Ni-Mo
Support	8 wt% SiO_2	10 wt% TiO_2
MoO_3 wt%	14.5	12.9
NiO wt%	3.0	2.8
Pore Volume Distribution by Hg Porosimetry; Surface Area by N_2 BET		
Total PV, cc/g	0.75	0.57
PV > 250A cc/g	0.14	0.02
PV > 160A cc/g	0.18	0.05
PV < 160A cc/g	0.57	0.52
PV < 100A cc/g	0.46	0.05
PV 100-160A cc/g	0.11	0.47
PM at (dv/dD)max Å	76	136
PM (BET), Å	74, 92	NA
Surf. Area, m^2/g	251	180
HDS-MAT, $C_{0.5g}$, %	91	80
Metals Distribution by XPS Analysis		
$(Mo/Al)_{int}$	0.11	0.12
$(Ni/Al)_{int}$	0.012	0.009
Mo Gradient	1.5	4.4
Ni Gradient	1.2	2.3

The properties of the magnesia-silica-alumina catalysts are summarized in Table VII. SN-6273 has a total pore volume of 0.60 cc/g, a macroporosity of 0.06 cc/g, about 57% of pore volume in pores with diameter between 100 and 160 Å and a pore mode of 116 Å as measured by mercury porosimetry. SN-6273 has a HDS-MAT activity of 70 and a uniform Mo distribution (low Mo gradient).

TABLE VII

NiMo CATALYSTS ON MAGNESIA–ALUMINA SUPPORTS

Catalyst	SN-6273X	SN-6273	SN-6542X	SN-6542
Sample	Support	Catalyst	Support	Catalyst
Impreg. Sol'n		Ni-Mo-P 0.72% P		Ni-Mo H₂O₂
MgO wt%		6.5		8.0
MoO₃ wt%		15.5		15.2
NiO wt%		3.2		3.2
Pore Volume Distribution by Hg Porosimetry				
TPV, cc/g	0.86	0.62	0.83	0.65
MPD (Vol), Å	137	125	139	129
TSA (N₂), m²/g	198	158	NA	168
Metals Distribution by XPS Analysis				
$(Mo/Al)_{int}$		0.12		0.12
$(Ni/Al)_{int}$		0.012		0.017
Mo Gradient		1.7		5.2
Ni Gradient		0.8		2.4

The properties of three catalysts based on a lithia-alumina support are presented in Table VIII. SN-6611, SN-6612 and SN-6613 were prepared by dry impregnation of SN-6614 with three different types of impregnating solutions containing respectively citric acid, hydrogen peroxide, and phosphoric acid as

TABLE VIII

NiMo CATALYSTS ON LITHIA-ALUMINA SUPPORTS

Catalyst	SN-6611	SN-6612	SN-6613	SN-6614
Impreg. Soln'	Citric	H_2O_2	Ni-Mo-P	Support
Li_2O wt%	1.0	1.0	1.0	1.2
P wt%	0	0	0.7	0
MoO_3 wt%	15.2	15.0	15.1	0
NiO wt%	3.2	3.3	3.0	0
Total PV, cc/g	0.63	0.62	0.60	0.79
PM at (dv/dD)max Å	128	129	120	125
PM (BET), Å	15	116	108	114
Surf. Area, m^2/g	187	174	173	218
HDS-MAT, $C_{0.5g}$, %	78	80	85	NA
Metals Distribution by XPS Analysis				
$(Mo/Al)_{int}$	0.12	0.10	0.11	NA
$(Ni/Al)_{int}$	0.014	0.015	0.014	NA
Mo Gradient	1.1	4.1	1.3	NA
Ni Gradient	1.0	0.93	1.1	NA

stabilizers. All three catalysts have similar physical properties including having at least 75% of their pore volume in the range of 100-160 Å. Also these catalysts have comparable HDS-MAT activities with SN-6613 being slightly higher.

TABLE IX

BERTY RESID MILD HYDROCRACKING ACTIVITIES

Catalyst	650 F+ Conversion vol%	1000 F+ Conversion vol%	IP Sediment %	HDS Activity %
Catalyst A	29	78	0.7	69
Catalyst B	40	78	0.9	71
SN-6001	49	82	1.8	45
SN-6010	48	88	1.0	64
SN-6224	40	85	0.9	66
SN-6264	41	86	0.7	71
HDS-2443B	45	83	0.9	71
SN-6571	46	91	1.2	71
SN-6572	43	87	0.7	70
SN-6412	53	89	2.4	38
SN-6650	39	84	0.6	68
SN-6262	39	79	0.9	73
SN-6411	44	84	1.5	46
SN-6542	44	91	0.7	64
SN-6273	40	88	0.8	70
SN-6612	58	92	0.9	64

B. Berty Reactor Mild Hydrocracking Activities

The MHC activities of the catalysts were determined using a Berty continuous stirred tank reactor as described in the experimental section. Two commercial alumina based catalysts, Catalyst A and Catalyst B with bimodal pore structures are used as the reference in the evaluation for MHC activities. Catalyst A and Catalyst B were equally active for the 1000 F+ (538 C+)conversion, however, Catalyst B was more active for the 650 F+ (343 C+) conversion. Therefore, Catalyst A is used as the base catalyst for the comparison of conversion advantages summarized in Table IX.

1. Lithia-alumina based catalysts (SN-6612)

As shown in Table IX, SN-6612 has the greatest activity for both 650 F+ (343 C+) conversion and 1000 F+ (538 C+) conversion. Compared to Catalyst A, SN-6612 showed an increase in 650 F+ (343 C+) conversion by 29 vol%, or a 100% improvement in relative conversion. SN-6612 also gave an appreciable improvement in the 1000 F+ (538 C+) conversion (14 vol%). The IP sediment-make showed only a minimal increase of 0.2%.

2. Alumina based catalysts
(SN-6001, SN-6010, SN-6224, SN-6264)

The mild hydrocracking activities for four NiMo and NiMoP catalysts on alumina supports with controlled pore size distributions (shown in Table III) are compared in Table IX. The four catalysts exhibited up to 20 vol% improvement in 650 F+ (343 C+) conversion and up to 10 vol% improvement in 1000 F+ (538 C+) conversion activities. The conversion improvement is presumably due to the unique pore size distributions that allow more reactant molecules in the 1000 F+ (538 C+) fraction to enter the catalytically active pores.

SN-6001 was highly inoperable because it generated a very high sediment-make. In contrast, the NiMo counterpart, SN-6010, still caused some reactor unit plugging problems but yielded much lower sediment than the SN-6001. As evidenced by the results with samples SN-6264 and SN-6224, as the catalyst macroporosity was increased the unit operability improved significantly.

3. Zeolite-containing catalysts (SN-6571, SN-6572)

The SN-6571 catalyst showed very high activities for 650 F+ (343 C+) and 1000 F+ (538 C+) conversion. Unfortunately, it also caused plugging problems even though SN-6571 and SN-6224 are similar in pore size distribution. To improve the unit operability, Valfor CP300-56 USY zeolite in SN-6571 was replaced with a dealuminated Y zeolite (Valfor CP300-35 SUSY obtained from PQ Corp.) in the preparation of SN-6572. As shown in Table IX SN-6572 gives about 14 vol% advantage in 650 F+ (343 C+) conversion and about 9 vol% improvement in 1000 F+ (538 C+) conversion. Most importantly, the sediment-make of SN-6572 is maintained at the same level as Catalyst A.

No unit plugging problems were experienced for SN-6572. The results demonstrate that the combined modifications of zeolite and pore structure provide a solution to the unit operability problem that occurred on the USY zeolite/alumina based catalysts.

4. Silica-alumina based catalysts promoted with B, Mg, Si, Ti (SN-6412, SN-6273, SN-6542, SN-6650, SN-6262, SN-6411)

As seen in Table IX, the boria-silica-alumina catalyst, SN-6412, showed 24 vol% improvement in 650 F+ (343 C+) conversion and 10 vol% higher 1000 F+ (538 C+) conversion relative to Catalyst A. However, SN-6412 produced about three-fold greater sediment content than Catalyst A. Another boria-silica-alumina catalyst, SN-6650, showed advantages in both the 650 F+ (343 C+) and 1000 F+ (538 C+) conversions. The sediment production was reduced very significantly down to 0.6 wt% for SN-6650 from 2.4 wt% for SN-6412. The magnesia-silica-alumina based catalyst, SN-6273, exhibited 11 vol% advantage of 650 F+ (343 C+) conversion and 10 vol% improvement in 1000 F+ (538 C+) conversion. In contrast to SN-6412, SN-6273 maintained a sediment-make similar to Catalyst A. Another magnesia based catalyst, SN-6542, gave slightly better performance than SN-6273.

The silica-alumina based NiMoP catalyst, SN-6262, was considered unsatisfactory for this MHC process because it failed to give any improvement in the 1000 F+ (538 C+) conversion, and yet had a sediment-make higher than Catalyst A. The titania-silica-alumina based NiMo catalyst, SN-6411, was highly inoperable although it showed moderate improvement in the 650 F+ (343 C+) and 1000 F+ (538 C+) conversion activities. The high sediment formation may be attributed to the monomodal pore structure of SN-6411.

In summary, the lithia-alumina catalyst, SN-6612; two magnesia-alumina catalysts, SN-6542 and SN-6273; and the boria-alumina catalyst, SN-6412 outperform the rest of the catalysts. Compared to Catalyst A, SN-6542 gave about 12 vol% improvement in 1000 F+ (538 C+) conversion and SN-6273 showed about 10 vol% improvement. The SN-6412 also exhibited a degree of improvement similar to SN-6273.

Sediment generally decreases with increasing macroporosity. This trend indicates that sediment should decrease with increasing micropore diameter as well. Both magnesia-alumina catalysts not only gave 1000 F+ (538 C+) conversion improvement but also maintained sediment contents similar to

Catalyst A and Catalyst B. In contrast, the boria-alumina catalyst, SN-6412, showed about three-fold greater sediment content at about 10 vol% higher 1000 F+ (538 C+) conversion relative to Catalyst A.

CONCLUSIONS

The conventional hydrocracking catalysts that consist of acidic cracking components such as Y zeolite, though exhibiting conversion improvements over alumina based catalysts, were not suitable for processing of heavy oils in the mild hydrocracking mode because of high sediment formation. In contrast, alumina catalysts containing basic oxides (alkali metal and alkaline earth metal) not only improve heavy oil conversion but, also maintain the sediment make at the same level as alumina based catalysts. The sediment make generally decreased with increasing macroporosity.

REFERENCES

1. U.S. Patent 4,686,030
2. U.S. Patent 4,844,792
3. Europatent Patent Application 0 247 678
4. U.S. Patent 4,724,226
5. U.S. Patent 4,600,498

ACKNOWLEDGEMENTS
We would like to express our appreciation to American Cyanamid Inc. for providing the supports for these catalysts.

8 Residuum Upgrading by High Pressure Slurry Phase Technology

Technical, Economic and Environmental Aspects

Klaus Kretschmar and Fritz Wenzel

VEBA OEL Technologie und Automatisierung GmbH
45899 Gelsenkirchen, Johannastr. 2-8, Germany

INTRODUCTION

The technology of petroleum residue upgrading in slurry-phase reactors under high hydrogen partial pressure is based on the Bergius-Pier technology for coal liquefaction commmercialized in Germany since the mid-1930's.

The coal liquefaction plants that still existed on VEBA OEL's refinery site were reactivated in 1952 and used for residual oil conversion until economic reasons forced their shut-down in 1964. During this period more than 12 Mio t residual oils were converted into light distillates at conversion rates above 90 wt% based on 524°C+ disappearance. The single train capacity at this time was limited to approximately 20t/h vacuum residue feed so that several trains were operated in parallel.

The slurry-phase reactors showed an inner diameter of 0.8 m and a length of 18 m. Three reactors were operated in series at a total pressure of up to 300 bar. The inner insulated cold-wall reactors did not contain any internals and were quenched with cold recycle gas in order to adjust the required operating temperature. One to three percent of a finely ground lignite coke impregnated with metal compounds such as iron sulfate was mixed with the residue feed. Without the addition of this additive a stable operation of the slurry-phase reactors was not possible.

From the mid-1950's the slurry-phase reactor cascade was connected with a hot separator and the hot separator overheads directly routed to a fixed bed reactor. Through this mode of operation it was possible to perform the primary conversion of residual oil as well as the secondary hydrofinishing towards refinery sales products within a single step.

Based on this experience VEBA OEL AG reactivated this technology in 1978 on a bench-scale, pilot-scale and semi-commercial scale level. The main tasks were the adaptation of operating and process conditions to modern construction materials, process control and operating philosophy and product quality requirements.

The intensive discussion about liquid and solid waste utilization led to considerations about using this technology for that purpose. After several years of research and development activities a processing route is today available that not only allows the conversion of petroleum residues at conversion rates up to 95% into high quality distillates but also allows a commingled operation with organic wastes such as lubricants, solvents, chlorinated hydrocarbons, PCB's, paint sludges, activated carbon and mixed plastic material.

This has been demonstrated since 1987 in a plant with a capacity of 24 t/h (3500 BPD). From mid-1993, this plant is scheduled to convert 40,000 t/y plastic waste into synthetic crude oil.

PRINCIPLE OF SLURRY PHASE HYDROGENATION

The principle of a slurry-phase reactor is shown in Figure 1. The feed, consisting of residual oil, additive and hydrogen flows through the reactor from the bottom to the top. At temperatures in the range of 450-490°C and pressures

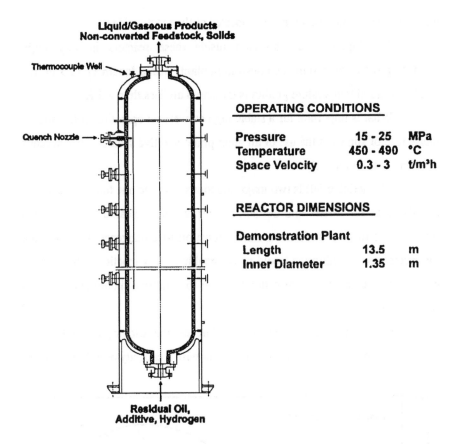

Fig. 1: Slurry Phase Hydrogenation Reactor

in the range of 15-25 MPa long chain molecules are thermally cracked and this
is followed by a reaction of the cracked material with hydrogen. The preferred
mode of operation employs a three reactor cascade.

At the outlet of the third reactor about 95% of the feedstock is
converted into distillates. At the same time a major part of the residual oil's
sulfur, nitrogen and oxygen are converted into hydrogen sulfide, ammonia and
water. Part of the hydrogenation enthalpy is used to preheat the reactor inlet
flow to operating temperature, while another part is quenched by cold recycle

gas injected through quench nozzles located at the reactor side.

The liquid-phase dispersion inside these reactors is very high. Regarding the reactor in the demonstration plant, with a length of 13.5 m, the axial and radial temperature gradients measured are less than 2-3 K.

What is important for a stable high conversion operation of the slurry-phase reactor shown is the use of an appropriate additive such as carbonaceous material, e.g., lignite coke.

The additive fulfils two major functions. On the one hand, it serves as an adsorbent for metals and destabilized asphaltenes released from the feed material. It thereby performs a carrier function to homogeneously distribute these compounds within the liquid phase, enable further reaction and transport out of the reactor. To perform this function the additive should show a high specific surface.

On the other hand, the additive influences the reactor phase hold-ups. In Figure 2, the start-up phase of a pilot plant with a capacity of 0.5 t/h vacuum

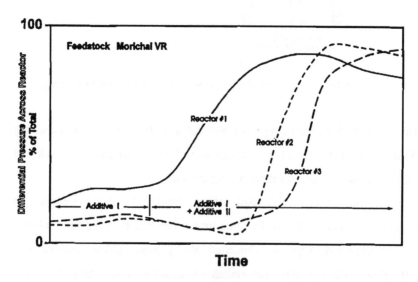

Fig. 2: Solid Influence on Slurry Hold-up in Hydrogenation Reactor

residue shows this effect. The reactors were brought on stream with vacuum residue. The feed was mixed with finely ground additive I and the temperature was increased to operating temperature. The static differential pressure measured across the cylindrical part of the reactors stays almost constant at a low level. This represents a relatively high gas hold-up. After start-up of a second additive feeding system with bigger particle sizes, the differential pressure steadily increases towards a constant upper limit. This value represents a gas hold-up of only a third of the previous value. By this effect the liquid phase residence time is increased more than twice, resulting in a significant increase in residue conversion and thereby release of process heat, which is compensated for by reduction of preheater outlet temperature and higher quench gas injection rate.

The additive particle size distribution must be tailored to the reactor geometry by applying a sedimentation/dispersion model, as was done in this example. The particle size of part of the additive must be adjusted in order to achieve a certain solid concentration inside the reactor by sedimentation.

REACTIVITY AND YIELD DISTRIBUTION IN SLURRY PHASE HYDROGENATION

For the determination of reactivity and yield distribution a wide variety of vacuum residues in different sized plants were tested. Beside residues from sweet, low metal crudes such as African Nigerian Medium or Amna with sulfur contents below 1% and Ni/V contents below 100 ppm, sour, high metal residues such as Venezuelan Boscan with a sulfur content of 6% and Ni/V content of 2000 ppm were also tested. Extensive test runs were also performed with Canadian Cold Lake with a sulfur content of 6% and Ni/V content of 450 ppm. Sulfur- and metal-rich Athabasca bitumen from the Canadian tar sand region of Alberta was also tested in high conversion operation. Due to the extraction process this residue contains up to 3% clay. It is also worth mentioning test runs with thermal or physical pretreated feedstocks such as residues from deasphalting units, slurry oils from FCC plants and vacuum residues from Visbreaker plants.

All the aforementioned feedstocks can be converted up to 95% into distillates by applying the slurry phase technology. In Figure 3, typical product

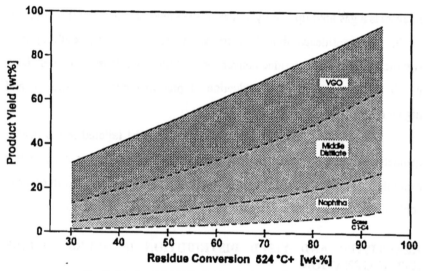

Fig. 3: Product Yield Distribution in Slurry Phase Hydrogenation

yields for C_1-C_4 hydrocarbons, naphtha, middle distillates and vacuum gasoil are given. At 95% conversion a C_5+ yield of approximately 85 wt% is achieved. The C_1-C_4 yield amounts to 10 wt% with a propane/butane portion of 4 wt%. According to the crude provenance oil specific differences from these typical yields occur. The most significant differences are observed within the C_1-C_4 yield. Nigerian Medium, for example, shows a C_1-C_4 yield of 7.7 wt% while Mexican Maya yields 12 wt% C_1-C_4 gases at 95% conversion.

As shown in Figure 4, the hydrogen consumption within the slurry

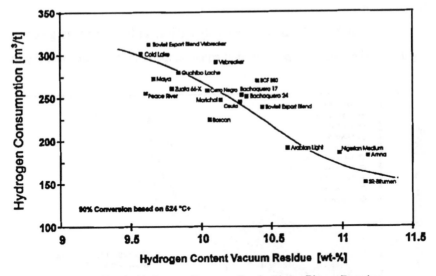

Fig. 4: Hydrogen Consumption in Slurry-Phase Reactor

phase hydrogenation is strongly dependent on feedstock quality. Paraffinic crudes such as Amna or Nigerian Medium show the highest hydrogen content and the lowest hydrogen consumption. With transition to naphthenic and heavy crudes the hydrogen consumption increases with decreased hydrogen content of the feedstock.

What is interesting in this context is the comparison of hydrogen consumption for Soviet Export Blend with 250 m^3/t and the Visbreaker vacuum residue of this crude with 320 m^3/t. Through preprocessing in a Visbreaker the volumetric flow for a high conversion plant can be reduced by 20-30% but, compared to a straight run residue, this residue is less reactive and shows a higher hydrogen consumption and a significant higher C_1-C_4 gas yield. Considering further that Visbreaker distillates need additional hydrotreating the question arises as to whether it is advisable to combine such a carbon rejection process with a hydrogen addition process or not. Although this question must be analyzed carefully, in general there are no advantages for this combination.

For different feedstocks reactivity against sulfur content is given in Figure 5. The tests were performed at equal temperature and space velocity in

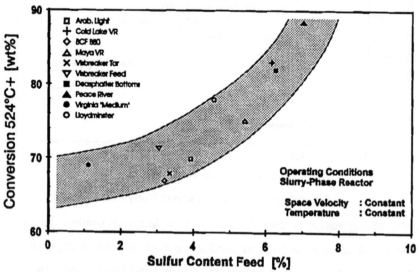

Fig. 5: Reactivity of Vacuum Residue versus Sulfur Content

the slurry phase reactor. Within a relatively wide band residue reactivity increases with sulfur content. Therefore, for low sulfur crudes reactor design has to be based on a somewhat higher temperature or lower space velocity to ensure high conversion. When designing a conversion plant for future feedstocks this will have to be carefully incorporated into the overall design.

OPERATING PRESSURE OF SLURRY PHASE HYDROGENATION
Closely connected with the design of a conversion plant for changing feedstocks is the question of the necessary pressure within the slurry phase hydrogenation. In Figure 1, the operating pressure was given as 15-25 MPa. The minimal necessary pressure at high conversion is dependent on the

feedstock characteristic.   A too low hydrogen partial pressure yields an increased solid formation and an insufficient asphaltene conversion.   This results in destabilization of asphaltenes in the slurry phase followed by deposition of asphaltenes on the reactor wall.   Lack of hydrogen through distortion of mass transfer results in further condensation and build-up of a clinging coke layer.

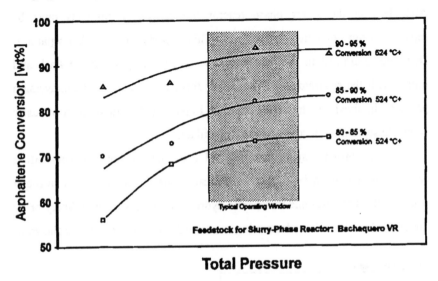

Total Pressure

Fig. 6: Asphaltene Conversion versus Process Pressure

In Figure 6, the dependence of asphaltene conversion on process pressure is given for Bachaquero feedstock.   At lower pressures a significant reduction of asphaltene conversion compared to the overall conversion is observed.   In an extreme case the destabilization of asphaltenes thus caused can be seen in the slurry-phase liquid product as a fluffy precipitation.   Such an operating condition has to be avoided.

At higher pressures all feedstocks tested so far could be converted at 95% in stable operation.   Some of the feedstocks such as Venezuelan Morichal can be processed at a pressure as low as 15 MPa without problems. For refineries with changing feedstocks it is very important to choose a process pressure adapted to the crude slate. If a conversion plant is to be designed for a specific feedstock it is sensible to optimize the process pressure.

INTEGRATED HYDROTREATING

After leaving the slurry phase reactor the conversion products must be separated from non-converted residue. This is performed in a hot-separator. Distillates entrained in the residue can be recovered by a vacuum flash and mixed with the separator overheads. For further treatment of the conversion products in fixed-bed hydrotreaters it is important that there be a virtually complete separation of solids from the distillates.

The primary synthetic crude oil does not contain any non-boiling components but must be further hydrotreated due to its sulfur and nitrogen content. This can be done in separate hydrotreating units in the refinery or by direct combination of a hydrotreating reactor with the hot separator as shown in

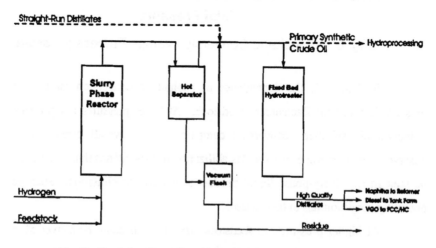

Fig. 7: Plant Configuration Slurry Phase Hydrogenation

Figure 7. The advantage of this integrated hydrotreating is that the high hydrogen partial pressure yields high quality products and that by just adding a fixed bed reactor a separate hydrotreating unit can be dispensed with, with the corresponding economic advantages.

The hydrogen consumption of the integrated hydrotreating is dependent on the severity of operation ranging from a mild hydrotreatment to a mild hydrocracking. Accordingly, there results a conversion of vacuum gasoil to naphtha and middle distillate, as shown in Figure 8. The typical operation of

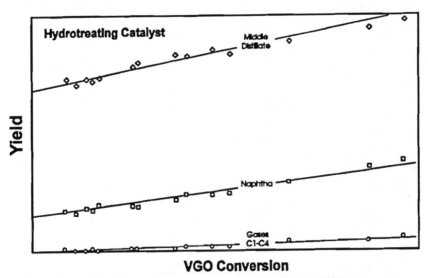

VGO Conversion

Fig. 8: Product Yield Integrated Hydrotreating versus VGO Conversion

the integrated hydrotreating is in the range of 20-40% VGO conversion, with a third converting into the naphtha boiling range and two-thirds into the middle distillate boiling range. Only minor amounts of C_1-C_4 hydrocarbons are released.

In Figure 9, the product qualities for the primary conversion products are compared with those from integrated hydrotreatment. The heavy naphtha at high severity operation shows sulfur and nitrogen contents of less than 1 ppm and can be routed directly to a reformer. The middle distillate shows a sulfur content of only 5 ppm and constitutes a high quality blending component. Due to its low sulfur and nitrogen content, as well as the absence of metals, the vacuum gasoil is an excellent feedstock for FCC- or HC-plants.

Figure 9: *Slurry Phase Hydrogenation Product quality with/without Integrated Hydrotreating*
- Feedstock: Arabian Light Vacuum Residue -

		Primary SCO	High Quality SCO
Light Naphtha, <65°C:			
Density, 15°C	g/ml	0.663	0.660
Sulfur	wt-ppm	760	<1
Nitrogen	wt-ppm	125	<1
Heavy Naphtha, 65-180°C:			
Density, 15°C	g/ml	0.756	0.752
Sulfur	wt-ppm	1,000	<1
Nitrogen	wt-ppm	520	<1
Bromine Number	$gBR_2/100g$	28.4	<1
Aromatic C	wt%	13	6
Middle Distillate, 180-350°C:			
Density, 15°C	g/ml	0.860	0.846
Sulfur	wt-ppm	13,200	5
Nitrogen	wt-ppm	1,400	14
Cloud Point	°C	-20	-22
Pour Point	°C	-25	<-15
Cetane Index		46	50
Vacuum Gasoil, 350-524°C:			
Density, 80°C	g/ml	0.913	0.886
Sulfur	wt-ppm	21,700	11
Nitrogen	wt-ppm	2,500	60
Pour Point	°C	19	25
Aromatic C	wt%	33	11
CCR	wt%	0.81	0.02
Metals (Ni+V)	wt-ppm	<1	<1

The volumetric product yield based on C_5+ is typically in the range of 103-106 vol%.

PROCESSING OF LIQUID AND SOLID WASTES

Apart from applying the slurry phase hydrogenation for residual oil conversion from crude oil processing, this process route can be used for processing of liquid and solid wastes commingled with residual oils. Lubricants, solvents, chlorinated hydrocarbons, PCB's, paint sludges, activated carbon and plastic wastes thus spent are converted into high quality synthetic crude oil. In the demonstration plant 6-7 t/h organic wastes are processed together with 18 t/h residual oils. In this plant it is possible to process about 1 t/h pure PCB.

From mid-1993 this plant will process 40,000 t/year mixed plastic waste from the "Duales System Deutschland". This involves plastics packaging separated from normal household trash which, by law, has to be recycled. The composition of typical plastic waste from German households is given in Figure 10. The main component is 60% polyethylene. An additional 5%

Figure 10: *Characterization of Mixed Plastics Waste from Households in Germany*

Type of Plastic:	PE	wt%	60
	PP	wt%	5
	PVC	wt%	10
	PS	wt%	15
	PA	wt%	5
	PET	wt%	5
Foll		wt%	42
Hollow Body		wt%	58
Elemental Analysis:	Carbon	wt%	78.1
	Hydrogen	wt%	11.4
	Sulfur	wt%	0
	Nitrogen	wt%	0.6
	Oxygen	wt%	2.3
	Org. Chlorine	wt%	5.6
	Ash	wt%	2

polypropylene, polyamide and polyethylenetherephthalate, 10% polyvinylchloride and 15% polystyrene can be found. The plastic waste is free of sulfur but, due to the PVC content, it shows 5.6% organic chlorine.

In tests mixtures of up to 30% plastic waste with vacuum residue were converted up to 95%. Depending on the type of plastic specific product yields and hydrogen consumptions are observed as indicated in Figure 11.

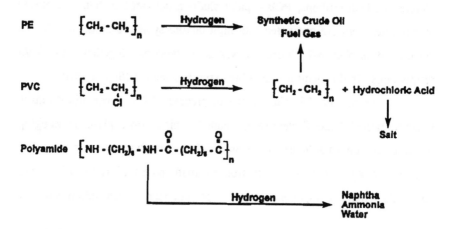

Fig. 11: Slurry Phase Hydrogenation of Plastic Waste

The polyolefines PE and PP show the highest liquid yield, approximately 93 wt% and the lowest hydrogen consumption of approximately 160 m^3/t. The resultant distillation curve of the synthetic crude oil is comparable with synthetic crude oil from residual oil processing. For polystyrene the liquid yield is similarly high but the hydrogen consumption at 7 wt% is higher due to its aromatic nature. The separation of hydrochloric acid from PVC is performed in a processing step prior to the slurry phase hydrogenation. Any remaining chlorine is split off inside the slurry-phase reactors, converted into a salt by the addition of a basic compound and released from the plant together with the hot separator bottom product. The hydrogen consumption of PVC is quite high, approximately 9 wt%. Polyamides show up nitrogen and oxygen in the polymere structure. These heteroatom bridges are split off almost completely, resulting in low boiling hydrocarbons depending on the amide type, as in this example hexane and decane. Oxygen and nitrogen are converted to water and ammonia, respectively. The hydrogen consumption for polyamides is approximately 6 wt%. PET's, whose polymer structure is bridged by oxygen, show a similar behaviour. Apart from hydrocarbons in the naphtha boiling range, some ethane and significant amounts of water, approximately 35 wt%, are produced. The hydrogen consumption is approximately 9 wt%.

In total, the slurry-phase hydrogenation of mixed plastic waste will result in a C_5+ yield of 88 wt%. Additional 3 wt% C_1-C_4 hydrocarbons, 4 wt% water and ammonia and 6 wt% hydrochloric acid will be produced. The hydrogen consumption will be approximately 330 m^3/t plastic waste.

This processing route of indirect material recycling can make a valuable contribution to future waste management concepts. The pollution problems related to thermal utilization or landfills of wastes do not occur with slurry-phase hydrogenation due to the closed mode of operation. Additional valuable energy resources can be saved.

ECONOMICS OF CONVERSION PROJECTS

The investment needed for a residual oil conversion project is considerable, although the differences between conventional technologies such as delayed coking and the newer hydrogenation technologies are not too big when compared on the basis of equal final product qualities. The economics of such a project can be increased by several measures, resulting in attractive rate of returns also with today's crude and product price structure. Besides increased processing of sour crudes and applying high conversion operation, the introduction of integrated hydrotreating and additional processing of liquid and solid wastes can contribute considerably to a profitable operation.

Assuming a refinery capacity of 10 Mio t/year and a conversion plant capacity on the basis of the slurry phase technology of 1.4 Mio t/year, additional revenues of $120 Mio/year result from switching to sour crudes on the basis of a price differential of $2/bbl for Arabian Light versus Arabian Heavy Crude.

The charges for the burning or landfill of organic wastes are going to increase significantly. In Germany today the costs are more than $600/t for burning and up to $300/t for landfills. Environmental legislation stipulates that waste must be avoided or recycled.

Assuming the blending of 25% organic wastes into the feed of a 1.4 Mio t/year upgrader at a charge of $450/t of waste, some $160 Mio/year in additional revenues will result.

By these measures the pay-out time of a conversion project can be reduced to below three years.

REFERENCES

1. Wenzel, F.; "Residual Oil Upgrading and Waste Processing in the VCC Demonstration Plant", 1992 NPRA Annual Meeting, New Orleans, USA.

2. Niemann, K.; "Status of the VCC Technology - An Update", 5th International Conference on Heavy Crude and Tar Sands", 1991 Caracas, Venezuela.

3. Kretschmar, K., et al.; "Process for the Hydrogenation of Heavy and Residual Oils", U.S. Patent 4,851,107.

4. Padamsey, R.; "Impact of Technology on the Economics of Upgrading Tar Sands Bitumen", 5th International Conference on Heavy Crude and Tar Sands, 1991 Caracas, Venezuela.

5. Kretschmar, K., et al.; "Primary Conversion of Heavy Oils and Bitumen to Synthetic Crudes. Construction and Operation ot the Pilot Plant. Final Report", 1988, ISBN 3-926994-0.

6. Deckwer, W.D.; "Reaktionstechnik in Blasensäulen", 1985, ISBN 3-7935-5540-2 (Salle).

9   Fate of Asphaltenes During Hydroprocessing of Heavy Petroleum Residues

A. STANISLAUS, M. ABSI-HALABI AND Z. KHAN

PETROLEUM TECHNOLOGY DEPT.; KUWAIT INSTITUTE FOR SCIENTIFIC RESEARCH, P.O.BOX 24885, 13109-SAFAT-KUWAIT.

ABSTRACT

Formation of coke like sediments or particulates is a serious problem in the hydroprocessing of heavy residues for high conversion. The sediments can cause both operability problems and rapid catalyst deactivation. The macromolecules of the heavy feedstocks such as asphaltenes are believed to contribute significantly to sediment formation and coke deposition. As part of an extensive research program on the factors which influence sludge or solids formation during residue hydroprocessing, we have examined the nature of changes that take place in the characteristics of the asphaltenic fraction of Kuwait vacuum residue under different operating conditions. The studies revealed that sediment formation is the result of reduction in solubilization efficiency of asphaltenes in the product medium compared with feedstock. Molecular size distribution of the product asphaltenes showed that operating at high temperatures enhanced depolymerization and fragmentation of asphaltenes to low molecular weight materials. A portion of the low molecular weight asphaltene fragments with relatively low H/C ratio resisted further cracking even at high temperatures and led to the formation of coke like sediments. Large pore catalysts were observed to reduce the problem of sediments formation. The role of catalyst pore size on asphaltenes conversion is discussed.

INTRODUCTION

Conversion of heavy petroleum residues to lighter cuts by hydroprocessing has gained considerable importance in recent years largely due to increasing market demand for cleaner, lighter and more valuable fuels or blending components. Catalysts consisting of molybdenum supported on γ-alumina and promoted by cobalt or nickel are extensively used in the process. The role of the catalyst is to promote the removal of sulfur, nitrogen and metal contaminants present in the feed by hydrodesulfurization (HDS), hydrodenitrogenation (HDN) and hydrodemetallization (HDM) reactions together with the conversion of heavier molecules to lighter products by hydrocracking.

Formation of coke like sediments or particulates is a serious problem in the hydroprocessing of residues for high conversion (1, 2). The problem becomes particularly more important at high temperatures when the conversion to distillates is beyond a certain level (Ca. 50%). The sediments can cause both operability and rapid catalyst deactivation problems.

Despite its importance as a critical factor limiting the maximum conversion attainable in commercial hydroprocessing units, the problem of sediment formation has not received much attention and the mechanism of its formation is not fully understood (3). The macromolecules of the heavy feedstocks such as asphaltenes are generally believed to contribute significantly to sediment formation and coke deposition (4-5). As part of an extensive research program on the factors which influence sludge or sediment formation during residue hydroprocessing, we have investigated the nature of changes that take place in the characteristics of the asphaltenic fraction of Kuwait vacuum residue under different conversions. The influence of catalyst pore size on asphaltene conversion and sediment formation was also examined.

EXPERIMENTAL

A commercial Ni-Mo/γ-Al$_2$O$_3$ hydroprocessing catalyst in the form of extrudates was used for the temperature effect studies. The surface area and pore volume of the catalyst were 297 m^2/g and 0.69 ml/g. Pore size effect studies on asphaltene conversion were carried out using three Ni-Mo/γAl$_2$O$_3$ catalysts with different proportions of mesopore and macropore volumes. The catalysts were prepared in this laboratory in accordance with a procedure reported elsewhere (6).

Kuwait Vacuum residue was used as feedstock. A fixed bed reactor system equipped for handling heavy feeds was used for the studies. The reactor had a total volume of 220 ml with an internal diameter of 1.9 cm. 50 ml of the catalyst (diluted with equal amount of pyrex glass beads) was charged into the reactor. The reactor was loaded with 6 mm pyrex glass beads above and below the catalyst bed so that the catalyst remained at the center of the reactor. Thermocouples inserted into a thermowell at the center of the catalyst bed were used to monitor the reactor temperature at various points.

After loading the catalyst the system was purged with nitrogen at a flow rate of 1 l/min. The temperature was increased to 523K (250°C) at a rate of 50°/h, while maintaining the nitrogen flow rate at 1 l/min. These conditions were maintained for 2 hours. The system was purged with hydrogen at a flow rate of 1 l/min for 30 min and then pressurized with hydrogen to 120 bars. Under these conditions, the presulfiding feed (recycle gas oil) was fed at a rate of 200 ml/h for 60 min.

When presulfiding was completed, the feed (Kuwait vacuum residue) was injected at 100 ml/h and the conditions were adjusted to desired operating temperature, pressure, hydrogen flow and LSHV. Testing was carried out under the following conditions: pressure, 120 bar; LHSV, 2h^{-1}; H$_2$/oil, 1000 ml/ml/h; temperature, 653-723K (380-450°C). After 6 hours of operation under the set conditions, liquid product samples were collected every 48 h for various tests.

TABLE 1. *Feed Characteristics*

Feed Property	Test Method	Values
Density @ 65°C (g/ml)	IP-190	0.9898
Density @ 15°C (g/ml)	IP-190	1.0224
API gravity (API)	D-1250	6.8
Sulfur (wt%)	XRF	5.2
Nitrogen (wt%)	Kjeldahl	0.4
CCR (wt%)	IP-13	15.7
Asphaltenes (wt%)	IP-143	6.5
Viscosity @ 100°C (Cst)	IP-71	788
Ash Content (wt%)	IP-4	0.04
Nickel (ppm)	ICAP	21
Vanadium (ppm)	ICAP	89

Feed and product samples were analysed using standard procedures. The properties of the feedstock together with the test method are listed in Table 1.

Molecular weight distribution were determined by gel permeation chromatography (Waters Associates).

RESULTS AND DISCUSSION

Effect of temperature on the conversion and characteristics of asphaltenes.

Table 2 summarizes the influence of operating temperature on the 797K (524°C) plus residue conversion asphaltene conversion and formation of toluene insoluble sediments. A remarkable increase in asphaltene conversion

TABLE 2. *Effect of Reactor Temperature on Asphaltene Conversion, 524°C plus Residue Conversion and Toulene Insoluble Sediments Formation.*

Temperature (K)	Asphaltene Conversion (wt %)	797K (524°C) plus residue conversion (wt %)	Toluene Insoluble Sediments (wt %)
653	22	19	0
683	25	25	0
703	27	51	0.2
723	42	64	3.0

is seen at temperatures above 703K with toulene insoluble materials starting to become significant at the same temperature. The yield of 797K (524°C) minus distillates is also increased with increasing temperature, the increase being more significant in the temperature region 683-703K.

The asphaltene and maltene fractions in the liquid products from various runs were separated and characterized by different techniques with a view to understand the fate of asphaltenes during the hydroprocessing of residues and their role in the formation of sediments. The atomic ratios of H/C in product asphaltenes and maltenes are plotted as a function of reaction temperature in Figure 1.

A substantial increase in H/C of asphaltenes is noticed when the reaction temperature is increased from 653-683K. Further increase of temperature to 703K results in a remarkable decrease in the H/C. A similar trend is also noticed in the H/C ratio of the non-asphaltenic (maltenes) fractions. However, the increase in the H/C ratio in the temperature range 653-683K is remarkably higher and the decline at temperatures above 703K is less

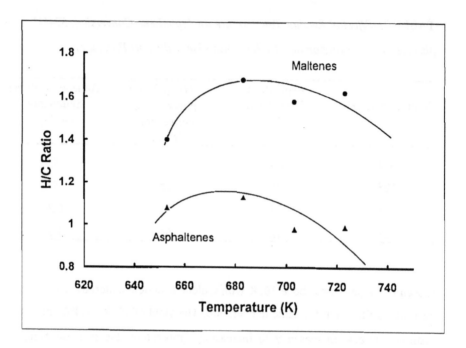

FIGURE 1. *Hydrogen/carbon (H/C) ratio of asphaltenes and maltenes vs operating temperature during hydroprocessing of Kuwait vacuum residue.*

pronounced for maltenes compared with asphaltenes. The sulfur and the vanadium contents of asphaltenés decrease steadily as the reactor temperature is increased (Fig. 2), but no such trends are noticed for nitrogen and nickel in the asphaltenes. The nitrogen and nickel contents of asphaltenes show an increase upto 703K (430°C) and then decrease with further increase in temperature to 723K (Fig. 3).

The average molecular weight of asphaltenes remaining in the product is found to decrease progressively with increasing temperature (Fig. 4). It is interesting to note that the decline in average molecular weight of asphaltenes is more pronounced than that of the maltenes.

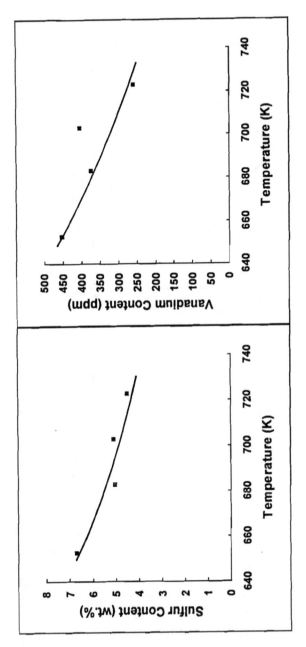

FIGURE 2. *Sulfur and vanadium contents of asphaltenes after catalytic hydroprocessing at different temperatures.*

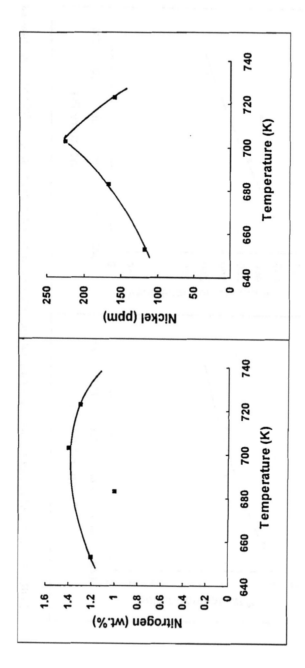

FIGURE 3. Nitrogen and nickel contents of asphaltenes after catalytic hydroprocessing at different temperatures

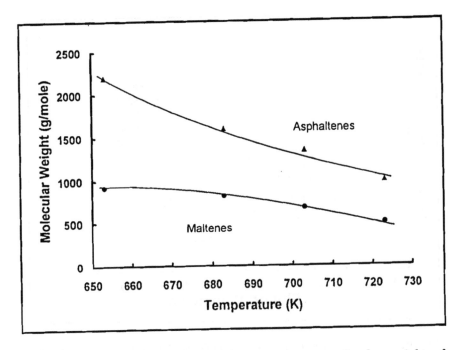

FIGURE 4. *Influence of operating temperature on molecular weight of asphaltenes and maltenes.*

Molecular weight distribution of product asphaltenes indicate a progressive shift in molecular weight to lower range with increasing temperature (Fig. 5).

Some explanation of these changes in the characteristics of asphaltenes can be advanced in terms of the molecular nature of petroleum asphaltenes. Asphaltene by definition is not a single molecular species but a solubility class of oil compounds, namely, the fraction of the oil that precipitates when mixed with a light straight chain aliphatic hydrocarbon such as pentane or heptane. It is believed that asphaltenes consist of sheets, held together to form large molecular entities or molecular aggregates [7, 8, 9]. The sheets contain large central polynuclear aromatic systems to which alkyl side chains or naphthenic rings are joined. The alkyl side chains and the naphthenic rings which contain

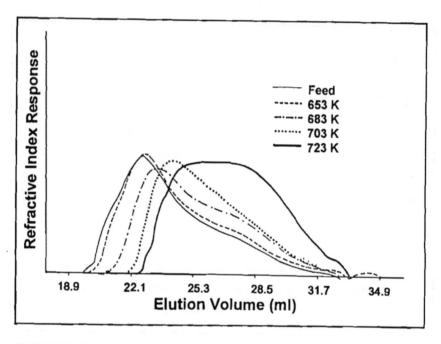

FIGURE 5. *Molecular weight distribution of asphaltenes ,in feed and products.*

sulfur bonds, aliphatic bridges and metalloporphyrin structures link the aromatic sheets. Association of the smaller molecular entities into larger species or molecular aggregates with special micellar properties is highly probable. Variation in molecular weight have been reported between 2850 (10) and 140000 (9).

During hydroprocessing of the residues, depolymerization and destruction of the asphaltene micelle may occur (11, 12). The observed pattern in the molecular weight distribution of asphaltenes in the products (Fig. 5) indicate that at lower temperatures (e.g. 653K the asphaltenes undergo no appreciable change. The reduction in molecular weight is substantially higher at higher temperatures. At temperatures in the range 683K –703K, the asphaltenes may depolymerize by cleavage of the relatively weak cross-linkage involving sulphur, aliphatic and metalloporphyrin bonds between the individual aromatic sheets which form the central part of an asphaltene molecule. Both depolymerization and destruction of micellar agglomerates by removal of sulphur and vanadium may take place simultaneously without attacking the individual aromatic sheets in this temperature range. Cracking the non-asphaltenic fractions is also likely to occur under these conditions. This is consistent with the observed decrease in S and V concentrations and the increase in the 797K (524°C) minus distillate yield. As the temperature is increased further to 723K, individual aromatic sheets of the asphaltene molecules appear to be attacked as indicated by further reduction in molecular weight, significant increase in asphaltene conversion and a substantial increase in the distillate yield.

The observed trend in the H/C ratio which first increases and then decreases with temperature (Fig. 1) is also consistent with the above explanation. It appears that separation of some individual sheets coupled to hydrogenation is taking place in the temperature range 653-683K leading to an increase in the H/C ratio. Above the latter temperature, scission of alkyl side chains and cracking of naphthenic rings from the individual sheets becomes important leaving the aromatic structures unaffected. The residual asphaltene appears to be more aromatic in nature as indicated by a lowering of H/C ratio. These highly aromatic and more refractory species may be the coke precursors. Toluene insoluble coke like sediments in the product are found to increase appreciably at higher temperatures. It is likely that the highly aromatic refractory portion of the asphaltenes which resist further cracking become incompatible in the oil (hydrotreated product) and precipitate out as sediments.

Effect of Catalyst Pore Size

One of the objectives of the present study is to examine the influence of catalyst pore size on asphaltene conversion and sediment formation during hydroprocessing of Kuwait vacuum residue. Three Ni-Mo/γ-Al$_2$O$_3$ catalysts (X, Y and Z) with different pore size distribution were used in the study. Tables 3 and 4 summarize meso (30-500Å) and macro (>500Å diameter) pore volume distribution data for the three catalysts. The micropores were not measured in this study.

Catalyst X is a mesoporous catalyst with no macropores. Furthermore, a major portion (about 60%) of the pore volume is present in pores having diameter between 100 and 250Å in this catalyst. Catalyst Y is essentially a macroporous catalyst with 58% of the pore volume contained in pores having diameter >500Å. More specifically about 43% of the pore volume is present in the 1000 - 3000 Å diameter pores of the catalyst. Catalyst Z contains both meso and macropores with about 70% of pore volume contributed by mesopores. A large percentage of the mesopore volume is contained in very narrow mesopores (30-100Å) in catalyst Z compared to catalyst Y.

TABLE 3. *Mesopore Size Distribution in Catalysts X, Y and Z.*

Catalyst.	Total Pore Volume (ml/g)	Mesopore Volume (ml/g)	Mesopore Size Distribution (%)		
			30-100A	100-250A°	250-500A°
X	0.53	0.53	38	60.5	1.5
Y	0.60	0.25	3.7	10.8	27.5
Z	0.69	0.48	37.8	21.7	10.1

TABLE 4. *Macropore Size Distribution in Catalysts X, Y and Z.*

Catalyst.	Total Pore Volume (ml/g)	Macropore Volume (ml/g)	Macropore Size Distribution (%)		
			500-1000A°	1000-3000A°	>3000A°
X	0.53	0	0	0	0
Y	0.60	0.35	15	43	0
Z	0.69	0.21	5.8	19.6	5.0

The activities of the three catalysts for asphaltene conversion and 797K (524°C) plus residue conversion as well as for the formation of toluene insoluble sediments are compared in Table 5. These data show that catalyst X with a large percentage of the medium size meso-pores (100-250 Å dia) is more effective for asphaltene conversion and for the cracking of residues with boiling range greater than 797K. Unfortunately, the formation of toluene-insoluble sediments is also relatively high for this catalyst.   Catalyst Z which contains

TABLE 5. *Effect of Catalyst Pore Size On Asphaltene Conversion and Sediment Formation.*

Catalyst	Asphaltene Conversion (wt %)	797K (524°C) Residue Conversion (wt %)	Toluene Insoluble (wt %)
X	57	52	0.58
Y	44	46	0.03
Z	48	50	0.13

mesopores both in the narrow (30-100Å) and intermediate (100-250Å) ranges together with some macropores ranks second for all conversions. The sediment formation is also reduced significantly for this catalyst. Catalyst Y which contains predominantly macropores together with some large mesopores (250-500Å dia) shows a comparatively lower activity for asphaltene conversion and residue cracking, but has the advantage of forming negligible amount of sediments. The results clearly show that catalyst pore size plays an important role in the formation of sediments during residue conversion. It would be possible to minimize the formation of undesirable sediments and improve various conversions by designing catalysts with optimum pore structure for residue hydroprocessing.

REFERENCES

1. W. I. Beaton and R. J. Bertolacini, Catal. Rev. Sci. Eng. 33, 281 (1991).

2. I. Mochida; X. Z. Zahao and K. Sakanishi, Ind. Eng. Chem. Res. 28, 418 (1989).

3. M. Absi-Halabi; A. Stanislaus and D. L. Trimm, Appl. Catal. 72, 193 (1991).

4. S. Komatsu; Y. Hori and S. Shimizu, Hydrocarbon Processing (May 1985) p. 42.

5. K. Saito and S. Shimizu. PETROTECH 8, 54 (1985).

6. M. Absi-Halabi, A. Stanislaus and H. Al-Zaid, Stud. Surf. Sci. Catal. 63, 155 (1991).

7. J. P. Dickie and T. F. Yen, Anal. Chem. 39, 1847 (1967).

8. C. Takeuchi; Y. Fukul; M. Nakamura; and Y. Shiroto, Ind. Eng. Chem. Process Des. Dev. 22, 236 (1983).

9. J. G. Speight and S. E. Maschopedis. "On Molecular Nature of Petroleum Asphaltenes ". In Chemistry of Asphaltenes (J. W. Bunger and N. C. Li, Eds), Advances in Chemistry Series, Vol. 195, p. 1, 1981.

10. J. G. Speight, Preprints, ACS Division of Petroleum Chemistry 32, 413 (1987).

11. S. Asaoka., S. Nakata., Y. Shiroto; and C. Takeuchi, Ind. Eng. Chem. Process Des. Dev. 22, 242 (1983).

12. Y. Shiroto, S. Nakata, Y. Fukul and C. Takeuchi, Ind. Eng. Chem. Process Des. Dev. 22, 248 (1983).

10   Catalyst Poisoning During Tar-Sands Bitumen Upgrading

J. D. Carruthers, J. S. Brinen, D. A. Komar, S. Greenhouse

CYTEC Industries, A Business Unit of American Cyanamid Company
CYTEC Research and Development
1937 West Main Street
Stamford, CT 06904

ABSTRACT

A number of hydrotreating catalysts are used in commercial heavy oil upgrading facilities. One of these, a $CoO/MoO_3/Al_2O_3$ catalyst has been evaluated in a pilot plant CSTR for Tar-Sands Bitumen upgrading. Following its use in a test of 200 hours duration, the catalyst was removed, de-oiled, regenerated by air-calcination to remove the coke, and then re-tested. Samples of the coked, fresh and regenerated catalyst were each examined using surface analytical techniques. ESCA and SIMS analysis of the coked and regenerated catalyst samples show, as expected, significant contamination of the catalyst with Ni and V. In addition, the SIMS analysis clearly reveals that the edges of the catalyst pellets are rich in Ca, Mg and Fe while the Ni, V and coke are evenly distributed. Regeneration of the catalyst by calcination removes the carbonaceous material but appears not to change the distribution of the metal contaminants. Retesting of the regenerated catalyst shows a performance similar to that of the fresh catalyst. These data serve to support the view that catalyst deactivation during early use is not due to the 'skin' of Ca and Mg on the pellets but rather via the poisoning of active sites by carbonaceous species.

175

INTRODUCTION

Major deposits of Tar Sands occur in Canada, estimated to be as large as 900 billion barrels of bitumen in place (1). Two distinct refining processes are used to convert the bitumen to crude oil:
    1. carbon rejection through Delayed or Fluid Coking, or
    2. hydrocracking via expanded bed catalytic hydrogenation (2).
This report describes a study of catalyst deactivation occurring during the latter upgrading route i.e. during hydrocracking via expanded bed catalytic hydrogenation.

Two projects have been completed in Canada during the past 5 years using this technology: at the Syncrude, Canada Upgrader operation in Fort McMurray, Alberta (LC-Fining, licensed by ABB-Lummus Crest, Cities Service and AMOCO), and at the Husky Oil Bi-Provincial Upgrader, Lloydminster, Saskatchewan (H-Oil, licensed by HRI Inc. and Texaco Development Corp.).

These processes hold out the most hope for increasing the yield of light products from bitumen without the formation of large quantities of low-value coke. The bitumen possesses heavy asphaltenic materials which, on heating, form species readily converted into coke. Catalytic hydrogenation can lower the concentration of such species and thereby shift more of the heavy material to lighter oils. Since both catalytic upgraders feed heavy process bottoms to cokers, to maximize light product yield it is important that catalytic hydrogenation minimize the concentration of those species which form coke (Conradson Carbon Residue).

Since 1988, a $CoO/MoO_3/Al_2O_3$ catalyst (base case) has been used at the Syncrude Bitumen Upgrader Expanded Bed Reactor in Canada to hydrocrack and hydrogenate bitumen feed. Target performance for this process is 65% conversion of the heavy material (distilling at temperatures above 525 deg.C) to lighter products along with high levels of desulfurization and CCR removal. It is now recognized that any improvements in catalyst activity must come from improvements in the ability of the catalyst to lower carbon residue (CCR) in the bottoms.

In addition to high levels of Conradson Carbon Residue, Tar Sands Bitumen from the Athabasca (Alberta, Canada) field is rich in contaminant metals, Ni, V and Fe. During the coking process these metals accumulate in the coke but during expanded bed hydrogenation processes, the metals build up on the catalyst.

Many workers have reported on the metal contaminants in Athabasca bitumen (3-8). Metals analyses for a typical sample of Tar Sands bitumen show the following:

V 212, Fe 160, Ni 78, Ca 114, Mg 12, Na 59 ppm (wt.).

It is generally believed that Ni, V and possibly Fe are present as porphyrins or chelated compounds associated with asphaltenes (5). But the presence of Al, Mg, Mn, Ca, and Ti is believed to be due to entrained mineral matter which is extremely difficult to separate from the asphaltenes (9,10). These metals deposit on the catalyst surface as sulfides causing active site poisoning and pore blockage (11) which accelerate the rapid deactivation of the catalyst. Since rapid deactivation is quite common for catalysts used in bitumen upgrading, it is important to decide whether coke deposition or metal contamination is the primary cause.

This report describes work carried out for Criterion Catalyst Co. L.P., during the development of improved catalysts for maximum CCR removal in expanded bed hydrogenation reactors. Criterion Catalyst Co. L.P., is a Joint Venture company (Shell-Cyanamid) conducting R&D at labs operated by the parent companies to generate new and improved catalysts and processes for the refining industry.

EXPERIMENTAL

Catalyst Samples

Much of the early work was carried out on several catalyst samples supplied by Syncrude, Canada, Ltd. following use in the commercial Expanded Bed LC-Finer at Ft. McMurray. These samples were de-oiled and subjected to analysis either in the coked or the calcined state. Later, catalyst samples were obtained following processing in a pilot scale CSTR reactor using Athabasca bitumen feed. These samples showed very similar features to those from the commercial unit.

Catalyst Activity Testing

Catalyst performance was evaluated using an Autoclave Engineers Continuous Stirred Tank Reactor (CSTR) with Robinson-Mahoney basket and internals, operating with once-through hydrogen at 1500 psi, 3500 scf/bbl treat rate and 1.0 LHSV (catalyst volume basis). The temperature was chosen to achieve 60% conversion of 525 deg.C+ (pitch). Feedstock properties are shown in Table 1.

Table 1   Feedstock Analyses

S, %	3.80
Ni, ppm	82
V, ppm	202
Fe, ppm	1000
MCR, %	14.2
Ash, %	1.00
sp. grav.	(1.00)
D1160	
IP, °F	555
10 vol %	697
20	784
30	853
40	913
975+, wt%	48.8

Surface Analysis

The first stage of the experimental program involved characterization of deactivated commercial catalysts obtained from the Syncrude LC-Finer. Samples of catalyst were examined using surface analytical techniques to determine the distribution of metals and other material which could act as catalyst poisons. The spent catalyst was treated with toluene in a soxhlet extractor to remove the oil prior to spectroscopic study. At this point, the catalyst was rich in coke. Catalysts in this state as well as catalysts where the coke was subsequently removed by calcination were examined. These measurements on commercial catalysts served to define the nature of the poisons as well as the experiments required. For the most definitive experiments, a similar catalyst was deactivated in the CSTR reactor using the Syncrude feedstock. This was treated as the catalyst described above to yield a coked and calcined pair. These catalysts were not as severely treated as the catalyst obtained from the commercial process but allowed controlled experiments under known conditions to be performed.

The two surface spectroscopic techniques employed in this study were X-ray Photoelectron Spectroscopy (XPS, also known as ESCA, Electron Spectroscopy for Chemical Analysis) and Secondary Ion Mass Spectrometry (SIMS).

ESCA spectra using a HP 5950A ESCA spectrometer (14), were obtained on the extrudate longitudinal axis to identify the surface composition and to compare coked and calcined spent catalyst surfaces. In some cases, whole extrudates and powdered extrudates were examined to compare surface and bulk compositions.

SIMS experiments were performed using a modified VG MICROLAB Mark II Surface Spectrometer which has previously been described in detail (14). The major feature of the SIMS instrumentation is a 30 kV gallium liquid metal ion gun which allows for the real-time display of secondary electron and secondary ion images generated by the ion gun. The VG data system SIMS software provides computer controlled data acquisition, data storage, and data manipulation in spectral, linescan, and imaging modes of operation.

In the SIMS experiment the surface is irradiated with a beam of ions (or atoms) and material is sputtered off the surface, resulting in the formation of positive ions, negative ions, and neutrals. These ions are detected using a mass spectrometer. SIMS by its nature is a destructive technique, i.e. the surface is continuously being eroded by the ion beam (sort of a chemical sandpaper) and is changing with time. Depending on the ion dose employed in the experiment, SIMS can be divided into two categories, static and dynamic. In static SIMS the ion dose is $<3 \times 10^{12}$ ions cm^{-2} (15). It is generally used for the analysis of organics and polymers since the low beam current density causes minimum disruption of chemical bonds thus resulting in the production of molecular species which are observed in the mass spectrum. The experiments described in this study were obtained at higher beam current densities (dynamic SIMS). Under these conditions, sufficient spectral intensity of elemental ions could be generated to allow chemical images and linescans to be obtained.

The SIMS linescan data were obtained using the gallium liquid metal ion gun operating at 12 kV with a beam current of 1 nA. The area scanned was approximately 4 mm^2. At this magnification the entire cross section of the catalyst extrudate pellet could be seen. The catalyst samples were prepared by cleaving them with a sharp knife blade normal to the central axis of the pellet. The pellet was then mounted onto a standard sample mount using a suspension of colloidal graphite in isopropanol. Usually three samples were mounted onto a single sample mount. In one case, the catalyst sample was prepared in a polymer matrix (typical SEM mount) and was polished.

The data were acquired using the SIMS linescan software. Since SIMS is extremely surface sensitive, the linescans show signal variations due to changes in the topography of the sample. To compensate for these variations, linescans were normalized using the Al m/z = 27 mass for positive ions and AlO m/z = 43 mass for the negative ions.

Chemical imaging is well established for SIMS. It can be used to obtain the spatial distribution of chemical components on the surface. The use of micro focusing liquid metal ion guns with beam diameters as small as 50 - 100 nm allows SIMS images to be obtained with sub-micron resolution. What makes SIMS most appealing is that it can be extremely sensitive to many ionic species, 10 - 100 times more sensitive than ESCA while achieving sub-micron spatial resolution.

SURFACE ANALYSIS RESULTS

ESCA

The ESCA spectra of all the catalysts examined in this study were very similar. Figure 1 shows a typical survey spectrum showing the major elements on the catalyst surface for a catalyst from a commercial operation after calcination to remove coke from the catalyst (Sample A, Table 2). Table 2 shows a compilation of the quantitative data derived from the survey scans for the catalysts investigated in this study. In the ESCA measurement, data are obtained from areas of six extrudate surfaces. The powdered catalyst represents contributions from a larger number of extrudates.

SIMS SPECTROSCOPY

SIMS spectra from the outer geometric surface obtained with the Ga ion gun showed the presence of metal ions and hydrocarbon species initially. Continued irradiation with the Ga ion gun resulted in an increased signal from

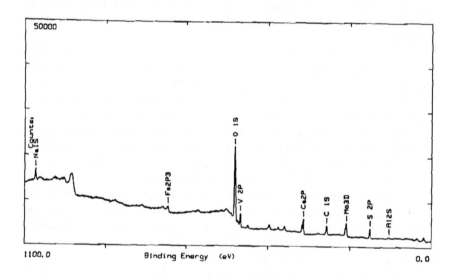

Figure 1   ESCA Survey Spectrum of the exterior surface of Catalyst A

Table 2   Summary of ESCA Data on Samples of "Used" Catalyst Atomic %

Catalyst:	A	A+	B	C	D	E	E+	F	G	G+
Na	2.0	1.5	(2.0)*	0.2	0.4	0.3	0.4	0.3	1.6	0.9
O	55.5	n.a.	40.0	9.3	17.4	14.9	n.a.	21.4	61.0	n.a.
V	1.8	11.7	0.1	0.1	0.2	0.1	4.6	0.3	0.5	6.3
N	n.d.	n.a.	n.d.	(1.6)	(1.5)	(1.0)	n.a.	(3.0)	n.d	n.a.
Ca	7.1	6.7	4.2	0.8	n.d.	1.7	0.3	0.3	2.5	0.4
C	18.4	n.a.	40.8	83.5	72.5	75.1	n.a.	62.9	8.5	n.a.
Mo	3.2	6.7	0.9	0.3	0.7	0.5	7.1	1.0	1.0	7.2
S	8.0	n.a.	5.7	1.8	2.0	2.4	n.a.	2.4	3.3	n.a.
Al	2.8	26.1	2.9	1.0	5.3	1.7	35.4	8.0	9.0	32.7
Si	n.d.	0.2	5.2	1.4	n.d.	2.3	0.1	0.4	11.3	0.02
Fe	1.1	0.4	0.3	n.d.	n.d.	n.d.	1.4	n.d.	1.0	1.9
Ni	n.d.	3.5	n.d.	n.d.	n.d.	n.d.	1.8	n.d.	0.1	2.3
Ti	n.d.	n.a.	n.d.	n.d.	n.d.	n.d.	n.a.	n.d.	0.4	n.a.

Catalysts:

A   Sample of catalyst from commercial expanded bed processing, calcined.

B   Sample of catalyst from commercial expanded bed processing, uncalcined.

C   Sample of catalyst from CSTR test #1, uncalcined.

D   Sample of catalyst from CSTR test #1, crushed, uncalcined.

E   Sample of catalyst from CSTR test #2, uncalcined.

F   Sample of catalyst from CSTR test #2, crushed, uncalcined.

G   Sample of catalyst from CSTR test #1, calcined.

Notes: In the table, ( ) is shown for N to indicate that the quantification is affected by the overlap with one of the Mo lines in this region. The concentration of Mo is also affected by overlap with the S2s line. The * next to the Na value indicates an estimate.

The data reported in columns identified with a '+' denote bulk chemical analyses for the calcined samples. Data marked as n.a. were not available.

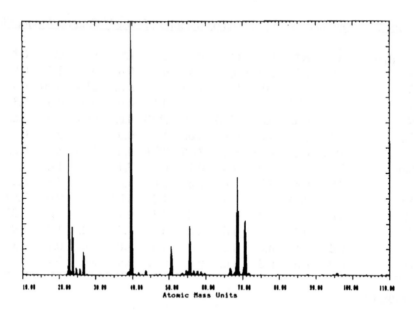

Figure 2   SIMS spectrum of the exterior surface of Catalyst A

Ca, Fig. 2 which becomes the dominant peak in the positive ion spectrum. The major peaks observed were from Na, Mg, Al, Ca, V, VO and Fe. Weaker peaks were observed for Mo, Ni and Co.

SIMS spectra were also obtained from a catalyst pellet which had been cleaved to expose its interior. Figure 3 shows a spectrum obtained from the interior cleaved surface. While the same elements are observed, the relative intensities of the peaks are dramatically different. Ca, Mg and Fe which are strong on the pellet exterior are weak on the internal surface while Al and V are stronger in the interior. In a similar fashion, SIMS spectra were obtained from a sample prepared in cross-section by embedding the catalyst pellet in an epoxy matrix and cutting and polishing a thin slice. The sample required sputter cleaning to remove contamination from the potting material. Similar spectra were obtained as discussed above, depending on whether the beam was positioned near the exterior or near the interior of the pellet. This sample was ideal for imaging experiments to map the distribution of the poisons throughout the catalyst.

In the negative ion mode, spectra obtained from the spent, uncalcined catalyst showed the presence of hydrocarbon fragments (C, CH, C_2H, C_2H_2, or CN) as well as O, OH, Cl and S.

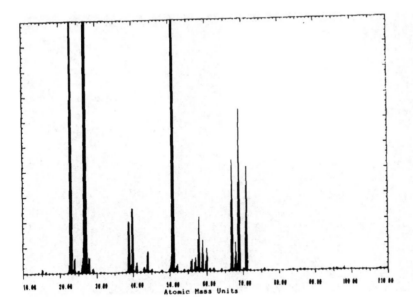

Atomic Mass Units

Figure 3   SIMS spectrum of the interior surface of Catalyst A

SIMS IMAGING

From the above spectral observations, a heterogeneous distribution of poisons was expected on the spent calcined catalyst. This was indeed observed in the imaging experiments. Figures 4 and 5 shows SIMS maps for Al, V, Ca, Na,

Mg and Mo obtained from Catalyst A. The Al map shows the outline of the catalyst cross-section clearly. Maps for Mo and V are similar to that of Al showing these metals to be relatively uniformly distributed across the catalyst cross-section. The Ca map shows that it is present primarily on the periphery of the pellet, i.e., on the skin. Mg and Na are also segregated towards the pellet exterior but are broader than Ca. SIMS maps were also obtained from the cleaved extrudate discussed above. Similar results were obtained showing a skin rich in Ca, Mg, Fe and Na. The imaging results from the uncalcined spent catalyst were very similar to those observed from the calcined catalyst. In addition to mapping the positive ion distribution, several of the negative ions were mapped for the uncalcined catalyst. For the negative ion mass fragments cited above, the distributions appeared uniform across the catalyst pellet.

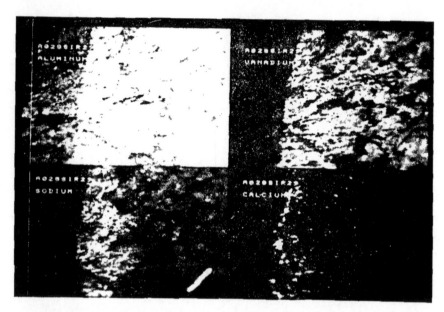

Figure 4   SIMS images showing distribution of Al, V, Ca and Na

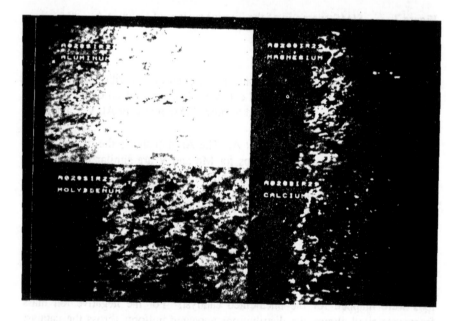

Figure 5   SIMS images showing distribution of Al, Mg, Mo and Ca

Table 3   CORRELATIONS

	Ca	Al	V	Mg	K	Mo	Fe	Co	Na	Ni
Ca	1.000	0.190	0.293	0.978	0.222	0.107	0.897	0.094	0.118	0.212
Al	0.190	1.000	0.941	0.130	0.957	0.964	0.254	0.905	0.968	0.945
V	0.293	0.941	1.000	0.237	0.965	0.936	0.351	0.881	0.923	0.981
Mg	0.978	0.130	0.237	1.000	0.162	0.051	0.852	0.047	0.060	0.160
K	0.222	0.957	0.965	0.162	1.000	0.954	0.296	0.884	0.955	0.962
Mo	0.107	0.964	0.936	0.051	0.954	1.000	0.160	0.945	0.971	0.962
Fe	0.897	0.254	0.351	0.852	0.296	0.160	1.000	0.159	0.199	0.281
Co	0.094	0.905	0.881	0.047	0.884	0.945	0.159	1.000	0.925	0.931
Na	0.118	0.968	0.923	0.060	0.955	0.971	0.199	0.925	1.000	0.946
Ni	0.212	0.945	0.981	0.160	0.962	0.962	0.281	0.931	0.946	1.000

SIMS LINESCANS

SIMS linescans across the catalyst samples, cleaved or cross-sectioned, were obtained and analyzed as described above. The linescans obtained for the calcined and uncalcined spent catalysts were essentially the same, with the exception of Na. The distributions of Ca, Mg and Fe were peaked at the pellet exterior, while the distributions of Al, Mo, V, Co and Ni were more uniform. The Na distribution was uniform in the uncalcined catalyst but was peaked at the pellet exterior in the calcined catalyst. A typical set of linescan data for the positive ions of interest are shown in Figure 6. A Table showing the correlation of the elemental distributions is given in Table 3. A value of 1 indicates a direct relationship, 0 no relationship. Negative ion linescans show these ions to be uniformly distributed across the catalyst pellet. The observation of a negative ion peak at 26 daltons might be indicative of the N distribution in the spent catalyst since the C_2H_2 peak is generally weak with respect to C_2.

The metals distribution of spent catalysts from the pilot plant CSTR reactor were very similar to those examined from the commercial operation. Primary emphasis was placed on the SIMS linescans which showed that Ca, Mg and Fe were peaked at the pellet exterior. Al, Mo, Co and Na were uniformly distributed across the catalyst pellet. K, Ni and V distributions were somewhere between the two extremes, showing a scalloped-shaped distribution. This distribution for Ni and V may be attributable to the relatively short period of time the catalyst was actually in the reactor relative to the commercial spent catalysts. It suggests that SIMS linescans could be used in a kinetic experiment to monitor the deposition of these poisons. A typical set of linescan data is shown in Figure 7.

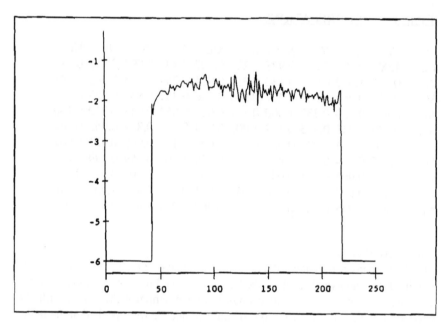

Figure 6a   SIMS Linescan for Molybdenum

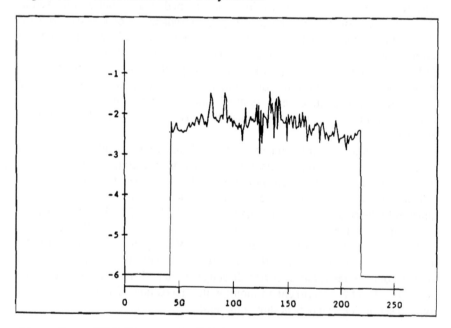

Figure 6b   SIMS Linescan for Cobalt

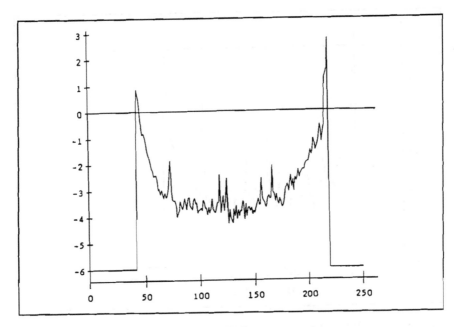

Figure 6c   SIMS Linescan for Magnesium

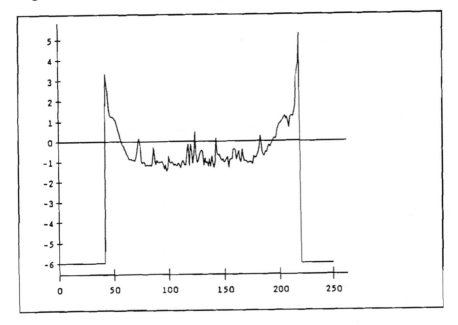

Figure 6d   SIMS Linescan for Calcium

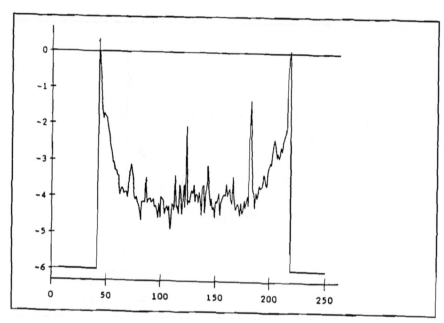

Figure 6e   SIMS Linescan for Sodium

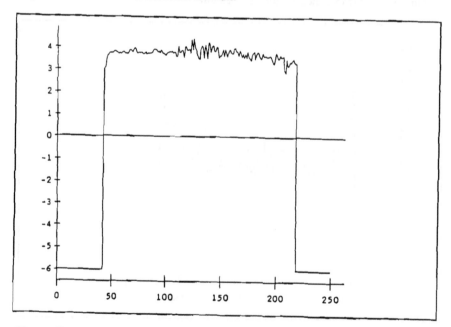

Figure 6f   SIMS Linescan for Iron

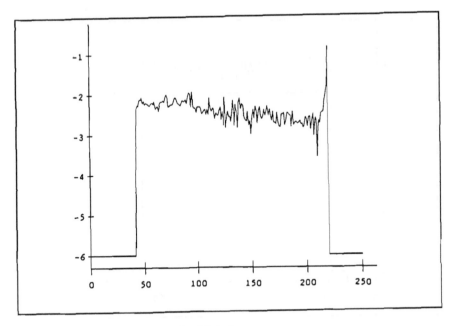

Figure 6g   SIMS Linescan for Nickel

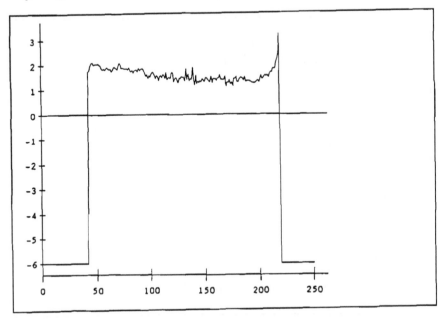

Figure 6h   SIMS Linescan for Vanadium

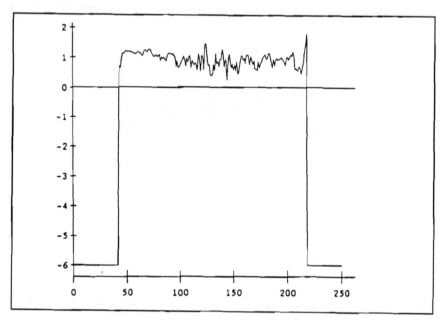

Figure 6i   SIMS Linescan for Potassium

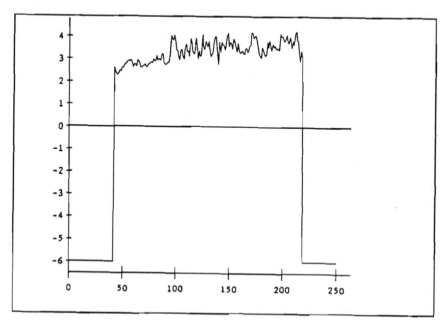

Figure 6j   SIMS Linescan for Carbon

The coincidence of the spikes in the Ca and Mg profiles is suggestive of the presence of a crystallite containing these elements. Negative ion distributions for the uncalcined catalyst were very similar to those discussed above. Calcination, at 600°C for 1 hour, did not significantly change the metals distribution.

Unlike ebullated bed catalysts used in petroleum oil hydroconversion, these catalysts used in the processing of Athabasca bitumen are now seen to exhibit high concentrations of calcium and magnesium on the external surface of the catalyst pellets. This is evident both for samples of considerable age removed from the commercial reactor as well as for those used for a short time in the pilot plant CSTR. It is plausible that these contaminants might trigger rapid early catalyst deactivation perhaps due to pore plugging or enhanced coke deposition in those same pores in the external 'skin' of the catalyst. Since calcination did not change the distribution of the calcium 'crust' on any used sample, the effect of this deposition might be estimated by catalyst activity testing of the 'before' and 'after' materials. The results of these tests are described in the next section.

CATALYST ACTIVITY TESTING RESULTS

Four catalyst tests are reported here. The first test involved three periods of increasing temperature (400C, 413C and 427C) with pitch conversions of 26%, 46% and 68% respectively. The second test involved constant temperature operation throughout the test period (418C) and pitch conversions near 60%.

The third test involved three temperature conditions (418C, 427C and 435C) but at a lower oil residence time in the reactor, generating pitch conversions of 44%, 52% and 58% respectively. At constant space velocity, the lower residence time was achieved by a larger catalyst charge. The charge was sufficient to allow catalyst to be removed from the reactor after 22 days on stream, de-oiled by Soxhlet extraction with toluene, and then regenerated by careful calcination in air. The fourth test evaluated the regenerated catalyst under the same conditions as test #2.

Under the conditions selected for this test program, each test was shortened, involuntarily, due to the inability to maintain the target oil temperature. The reactor contents are heated via a clam-shell jacket heater which provides heat through the reactor wall to maintain constant oil temperature and conversion. Several causes were proposed to account for this lack of ability to maintain oil temperature. All focussed on a sudden inability of the catalyst to continue supplying hydrogenation (exothermicity) activity to the system.

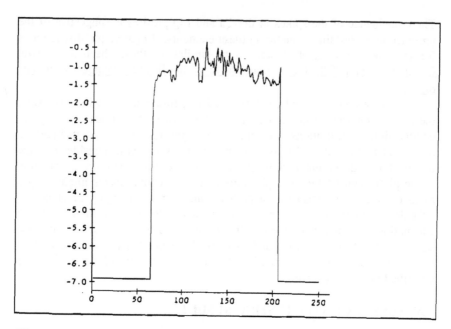

Figure 7a   SIMS Linescan for Molybdenum

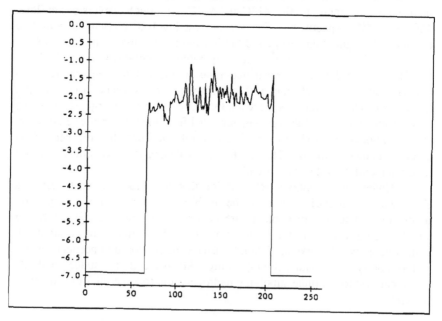

Figure 7b   SIMS Linescan for Cobalt

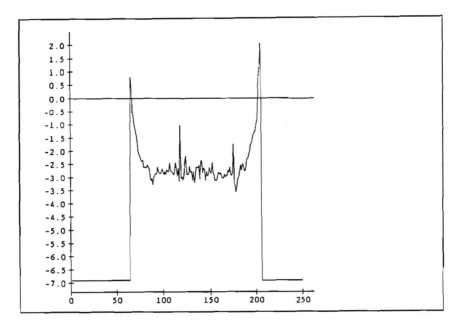

Figure 7c   SIMS Linescan for Magnesium

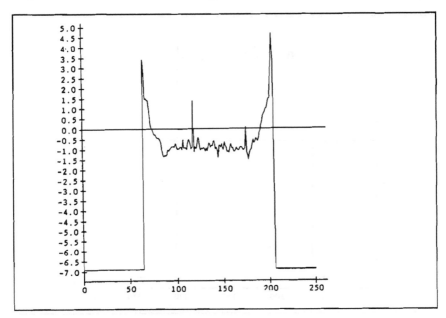

Figure 7d   SIMS Linescan for Calcium

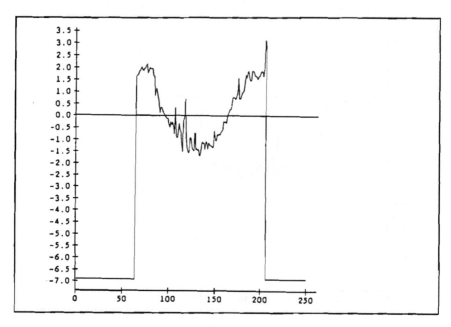

Figure 7e   SIMS Linescan for Potassium

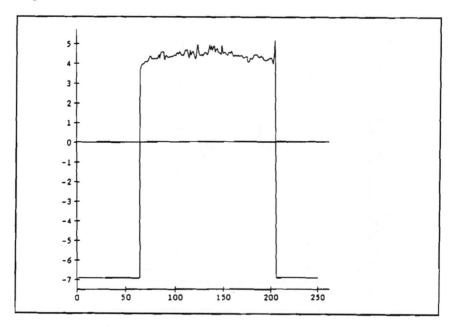

Figure 7f   SIMS Linescan for Sodium

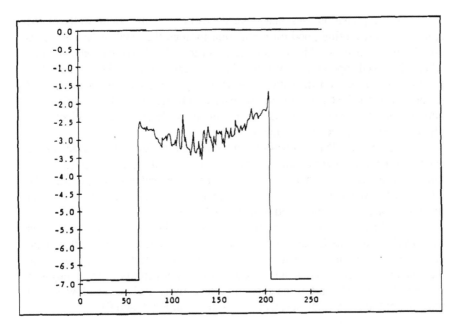

Figure 7g   SIMS Linescan for Nickel

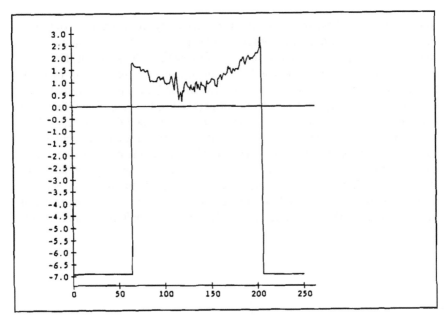

Figure 7h   SIMS Linescan for Vanadium

One explanation was that coke deposition on the inner walls of the reactor was preventing good oil circulation through the catalyst basket. This possibility was dismissed following a test, not included in the data shown here, which involved operation until the loss in exotherm could be detected, followed by a shut-down, catalyst wash in light oil, reactor clean-out and restart. The catalyst performance showed no change from its condition just before shut-down, indicating that the treatment, while clearing any blockages, did not reverse the decay in exothermicity.

A second explanation was that metal contamination of the surface of the catalyst pills was in some way inhibiting access of the oil molecules to the internal structure of the catalyst. By removing a used catalyst sample from the 22-day test (test #3) followed by regeneration and re-test, it was believed that this proposal could be supported or dismissed.

In Figs. 8 and 9, deactivation curves are shown for both sulfur and Microcarbon Residue (a substitute test for CCR - Ref.12) removal. It is quite clear that regeneration of a catalyst contaminated with metals from processing the Athabasca Bitumen does indeed restore the initial activity of the catalyst as depicted by the early part of the curve, but the activity declines to a lower level than that of the fresh catalyst (cf. curves A and B in Fig. 8 and 9).

DISCUSSION

SIMS analysis of these spent catalyst samples clearly show that the exterior of the catalyst pellets are rich in inorganic ions deposited from the bitumen. Ca, Mg and Fe are peaked at the exterior surface while others (V and Ni) more evenly distributed across the catalyst, with distributions resembling those of the catalytic elements. While this observation for Ca could be inferred from the ESCA data comparing powders with extrudates, Fig. 10, SIMS linescans and maps clearly show the distributions. In addition, these SIMS measurements show the uniform distribution of carbonaceous materials (coke formation) across the catalyst. The SIMS measurements also show that catalysts from the commercial process and from the pilot plant are very similar, differing only from the effects attributed to time on stream. Calcination of the spent catalyst removes much of the carbonaceous material but does not alter the distribution of the deposited inorganic matter (except for Na).

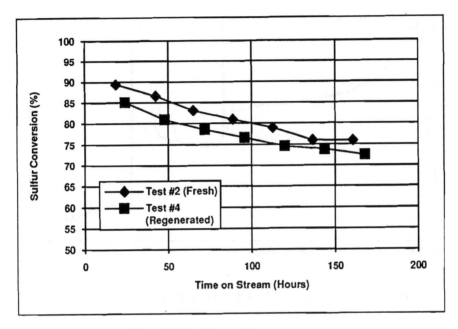

Figure 8   CSTR Test - Sulfur Conversion

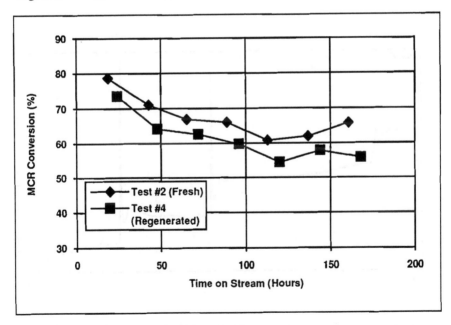

Figure 9   CSTR Test - MCR Conversion

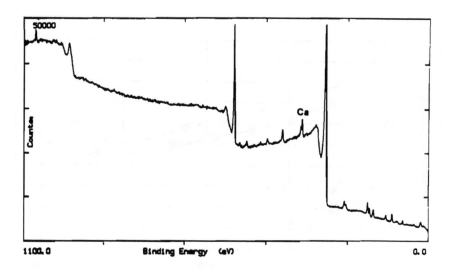

Figure 10a   ESCA Survey Spectrum of catalyst from CSTR

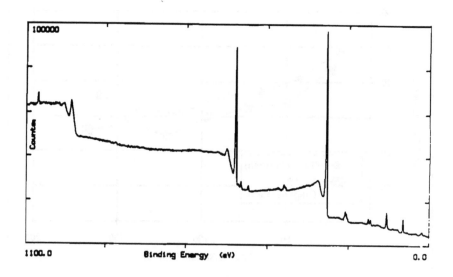

Figure 10b   ESCA Survey Spectrumof catalyst in Figure 10a, powdered

Catalyst performance testing of the calcined (regenerated) catalyst, however, shows clear evidence that the initial activity of the catalyst has not been damaged by the 'skin' of Ca and Mg deposits on the catalyst exterior surface. We may conclude, therefore, that the accumulation of contaminant metals at the catalyst external surface does not contribute markedly to the early stages of catalyst deactivation. Poisoning of active sites within the catalyst structure by contaminant metals has the effect of lowering the lined-out activity of the catalyst in much the same way as has been shown for petroleum residue processing catalyst (13).

Catalyst deactivation in the first 10 days of operation, much like that reported for petroleum residue processing, occurs through poisoning of active sites by adsorption of organic coke-precursors. In the specific case of Athabasca Tar Sands Bitumen processing, the carbonaceous deposits are distributed uniformly throughout the catalyst pellets and cause rapid pore volume loss, active site poisoning and massive loss of hydrogenation capability under the processing conditions described here.

CONCLUSIONS

Catalysts used in hydrocracking Athabasca Tar Sands Bitumen are found to deactivate suddenly when tested in a laboratory scale CSTR reactor under certain processing conditions. Contamination of the catalyst occurs from metals present in the bitumen. Calcium and Mg particularly, have been shown by surface analytical techniques to deposit preferentially on the outer 'skin' of the catalyst pellets while Ni, V, and Na are more uniformly distributed.

It can be inferred, that the Ca and Mg deposits do not cause rapid catalyst deactivation since 'used' catalysts, following de-oiling and calcination to remove the coke, return to almost full fresh activity on re-testing.

ACKNOWLEDGEMENTS

The authors wish to thank Syncrude, Canada, Ltd. and particularly W. Bishop and V. Nowlan for supplying both the samples of used commercial catalyst and the Athabasca Bitumen feedstock. Appreciation is also expressed to Cytec Industries/American Cyanamid Co. and Criterion Catalyst Co. L.P. for permission to present this paper for publication.

REFERENCES

1.   Berkowitz, N. and Speight, J.G., Fuel, 1975, 54, 138.

2.   Chrones, J. and Germain, R.R., Fuel Science & Technology
      International, 1989, 7 (5-6) 783.

3.   Budziak, C.J., Vargha-Butler, E.I., Hancock, R.G.V., Neumann, A.W.,
      Fuel, 1988, 67, 1633.

4.   Mossop, G.D., Science, 1980, 207, 145.

5.   Kotlyar, L.S., Ripmeester, J.A., Sparks, B.D., Woods, J.,Fuel, 1988, 67,
      1529.

6.   Yen, T.F., "The Role of Trace Metals in Petroleum," Ann Arbor Science
      Publishers, Michigan, 1975, p. 1-30.

7.   Jacobs, F.S., Filby, R.H., Anal. Chem., 1983, 55, 74.

8.   Hardin, A.H. and Teman, M., Reprints 2nd World Congress on Chem.
      Eng., 1983, 3, 134.

9.   Jacobs, F.S. Bashelor, F.W., Filby, R.H., in "Characterization of Heavy
      Crude Oils and Petroleum Residues," Symposium Internationale, Lyons,
      France, 25-27 June 1984, p. 173.

10.   Graham, W.R.M., Am. Chem. Soc. Div. Petr. Chem. Prepr., 1986, 31,
      608.

11.   Tamm, P.W., Harnsberger, H.F., Bridge, A.G., Ind. Eng. Chem.,
      Process Des. Dev., 1981, 20, 262.

12.   Annual Book of ASTM Standards, Method: D4530-85 Vol. 05.03, 1991,
      453.

13a. Sie, S.T., Studies in Surface Science and Catal. 6, Catalyst Deactivation,
      Elsevier, Amsterdam, 1980 p. 545.

13b. Dautzenberg, F.M., Van Klinken, J., Pronk, K.M.A,Sie, S.T., Wijffels,
      J.B., Proc. 5th Intern. Symp. on Chem. Reaction Eng., Houston, March

13-15, 1978; Chem. Eng. Sci., 1978, 254.
14.   Brinen, J.S., Greenhouse, S., Pinatti, L., Surface and Interface Anal.,1991, 17, 63.
15.   Benninghoven, A., Phys. Status Solidi, 1969, 35, K169., Briggs, D. and Hearn, N., Vacuum, 1986, 36. 1005., Hearn, N. and Briggs, D., Surface and Interface Anal., 1986, 9, 411.

11 Comparison of Unimodal Versus Bimodal Pore Catalysts in Residues Hydrotreating

M. ABSI-HALABI*, A. STANISLAUS AND H. AL-ZAID

PETROLEUM TECHNOLOGY DEPT., KUWAIT INSTITUTE FOR SCIENTIFIC RESEARCH, P. O. BOX 24885, 13109 SAFAT, KUWAIT.

ABSTRACT

Catalyst pore structure is a critical factor influencing the performance of residues hydroprocessing catalysts.   The effect is reflected in both hydrodesulfurization activity of the catalyst and its rate of deactivation. In this paper, the pore size distributions of two categories of catalysts, unimodal and bimodal, were systematically varied. Performance evaluation tests in a fixed bed reactor using vacuum residues under conditions comparable to typical refinery operations were conducted. Two series of unimodal and bimodal catalyst extrudates were prepared starting from boehmite gel, whereby the pore structure was systematically varied using hydrothermal treatment and organic additives. For the unimodal catalysts, the pore maxima ranged between 50 and 500Å with 70-80% of the pore volume in the desired pore diameter range. The bimodal catalysts had narrow pores with pore diameters less than 100 Å and wide pores with pore diameter around 5000Å. For bimodal catalyst, an increase in the average wide pore diameter, while maintaining the narrow pore constant, had no significant effect on the catalyst performance. For monomodal catalyst, the activity of the catalyst was noted to have an optimum between 150-350Å diameter. Furthermore, the performance of the catalyst concerning its desulfurization activity and deactivation was superior to that of the bimodal catalysts.

INTRODUCTION

Maximization of the throughput of heavy residues hydroprocessing units is anticipated to remain one of the objectives of petroleum refining over the foreseeable future. This stems from the impact of conversion on the profitability of refineries (1). The overall performance of an existing catalytic hydroprocessing unit is determined by the activity of the catalyst system and its deactivation rate.

In catalytic residues hydroprocessing, the key factors influencing the performance of the catalyst are the high metals and asphaltenic contents of the feedstock that lead to both coking and foulant metals deposition (2-4). Catalysts used in residue hydrotreating show in general initial rapid deactivation, followed by a more gradual activity decline, and, finally, accelerated aging (5). The period of initial deactivation has been attributed to coke deposition on the catalyst (2, 6, 7). Most of the coke is deposited within the first 24 h of operation.   Trimm et al. (8) proposed that the initial deactivation could be caused by partial surface poisoning by feed metals, in addition to coke deposition. The catalyst life is mainly determined by the intermediate deactivation period, generally attributed to gradual pore plugging by metal deposits (7, 9, 10), and eventually causing complete physical obstruction of the pores. Moreover, as the catalyst bed temperature is raised to maintain constant conversion in commercial operation, the increasing coke deposition at high temperatures can make an important contribution to the final aging (6, 9, 11).

Intensive research has been conducted over the past decade to develop improved catalysts for residues hydroprocessing and to determine the effects of various catalyst properties on performance. Due to the nature of the feedstock, there is a general agreement that the key property is the pore structure of the catalyst. The effects of catalyst pore size on hydrotreating of heavy oils and residues have been investigated by several workers(12-16). These studies involved both monomodal and bimodal catalysts and led to the conclusion that wider pore catalysts are more effective and more stable than narrow pore catalysts. However, most of these studies were conducted on catalysts from different sources; hence, interference by factors other than pore structure would influence the results and prevents reaching conclusive evidence.

This paper is part of a research program conducted in our labs to investigate various factors relating catalyst properties and preparation procedure with performance. The results of our systematic study of the effect of both monomodal and bimodal catalysts are presented. The catalysts were prepared from boehmite alumina through a special procedure developed in our labs to minimize the effects of other variables that may complicate the interpretation of the results.

EXPERIMENTAL

Boehmite alumina, SB-100, (Condea Chemie GmbH, Germany), carbon powder, CS-A4, (Norit, The Netherlands), and reagent grade ammonium heptamolybdate, aluminium nitrate, and nickel nitrate (Fluka, Switzerland) were used in the development of the catalysts. A sulfonated melamine formaldehyde resin (PLAST-1), developed in our labs(17), was used as plasticizing agent. A kneading machine model D-5277 (Linden, Germany) and a single screw type extruder model 250 (Netzsch, Germany) were used in making the extrudates. The surface area, pore size distribution, and side crushing strength of the catalysts were measured using Quantasorb adsorption unit (Quantachrome Corp., USA), mercury porosimeter model 9305 (Micromeritics, USA), and crushing strength equipment model PTB 300 (Pharma, France).

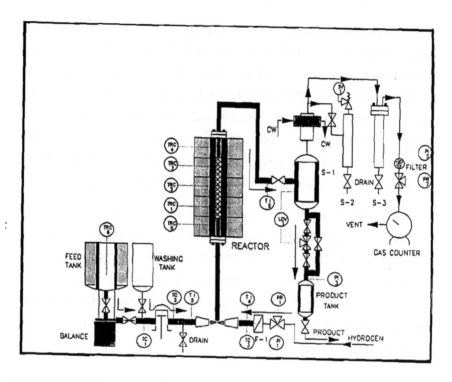

FIGURE 1. *Schematics of the pilot plant used in assessing the performance of the catalysts.*

The pilot plant used in the study was designed and constructed within Kuwait Institute for Scientific Research (Fig. 1). The feed section consists mainly of a heated feedstock tank and a metering pump. The feed flow rate is monitored by measuring the weight of the feedstock tank. The reactor is 75 cm long with 20 mm internal diameter and is equipped with a thermowell. The product section consists of a separator and a receiving vessel for the product.

The alumina extrudates used as support for the catalysts were prepared from boehmite alumina by kneading and peptizing with $HNO_3/AlNO_3$ solution in the presence of a plasticizing agent, PLAST-1, and other pore forming additives as required. The alumina paste was extruded through 1.5 mm orifice die, dried at 200 °C, and calcined at 550 °C. The pore structure of the monomodal alumina extrudates were modified by hydrothermal treatment according to the procedure reported by Absi-Halabi et al. (18,19) The extrudates were then impregnated with ammonium molybdate and nickel nitrate solutions, dried at 150 °C, and calcined at 550 °C.

The catalysts were first screened in a microreactor to determine their initial activity using vacuum gas oil as feedstock. Complete assessment of the catalysts was conducted on the pilot plant using vacuum residue obtained from Kuwait National Petroleum Co. The properties of the feedstock were determined using standard procedures and are listed in Table 1.

TABLE 1. *Physico-chemical Properties of the Vacuum Residue Feedstock.*

Property	Test Method	Unit	Feedstock
Density @ 15 C	IP-190	g/ml	0.9955
API Gravity	D-1250	API	10.6
Total Sulfur	XRF	wt %	5.2
Total Nitrogen		wt %	0.41
C.C.R.	IP-13	wt %	16.9
Kin-Viscosity @ 100°C	IP-71	cSt	840
Ash Content	IP-4	wt %	0.02
Metal in Ash			
Ni	ICAP	ppm	36.2
V	ICAP	ppm	78.9
Asphaltenes	IP-143	wt %	8.41

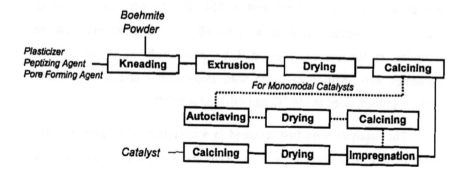

FIGURE 2. *Schematics of the preparation steps used in preparing the monomodal and bimodal catalysts.*

RESULTS AND DISCUSSION

To minimize complications in interpreting the effect of pore structure on the performance of residues hydrotreating catalysts, a preparative procedure was developed which permitted systematic variation of the pore structure (18, 19). The procedure is summarized schematically in Figure 2. Basically, boehmite alumina powder was kneaded in the presence of a peptizing agent and a plasticizer, extruded, dried and finally calcined. The alumina extrudates were then impregnated with Ni and Mo salts and recalcined to obtain the final catalysts.

For bimodal catalysts, wide pore-forming additives were mixed during the kneading stage and were combusted during the calcining step. High molecular weight carboxy- methylcellulose was found to be ideal in generating pores in the range 500-4000 Å. For pores larger than 5000 Å, carbon black, having an appropriate particle size distribution, was added. The percentage of wide pores in the catalyst was controlled by regulating the amount of pore forming additive.

TABLE 2. *Variation of the properties of the alumina extrudates with hydrothermal treatment conditions.*

Treatment Conditions		Alumina Extrudates Properties		
T, °C	*Time, h*	*Surface Area (m²/g)*	*Mean Pore Diameter(Å).*	*Crushing Strength(N/mm)*
150	1	193	80	10.5
	4	181	100	21
	8	164	150	17
200	1	180	120	12.5
	4	117	300	9.7
	8	97	410	7.2
300	1	23	800	5.3

For monomodal catalysts, γ-alumina extrudates with narrow monomodal pore structure were subjected to hydrothermal treatment in the presence of ammonia. Through this treatment, it is possible to widen the pores to the desired diameter by adjusting the temperature and treatment time. Table 2 shows the effect of treatment conditions on the key properties of the alumina extrudates, namely surface area, side crushing strength, and mean pore diameter.

For the purposes of the current study, four alumina extrudate formulations were selected for performance evaluation. Table 3 lists the characteristic preparation parameters for these formulations. The B series is bimodal and the M series is monomodal. The physical and chemical properties of the prepared catalysts are presented in Table 4. The chemical composition shows that all catalysts have almost identical composition, particularly with regard to the catalytically active components, Ni and Mo. The side crushing strength values were ensured to be reasonably high for industrial applications.

TABLE 3. *Composition and preparation procedure of alumina extrudates used for catalyst preparation.*

Catalyst Code	Alumina Formulation + Pretreatment
B1	Alumina, SB100/ Carbon black, CS-A4 (13 wt. %)
B2	SB100/High Molecular Wt. Carboxymethyl-cellulose (20 wt. %)
M1	SB100 (with no pore forming additives)
M2	Same as M1 but thermally treated with NH_3 at 200°C for 4 h.

TABLE 4. *Physical and Chemical Properties of the Prepared Catalysts.*

	Catalyst Property	B1	B2	M1	M2
I	Chemical Composition, (Wt % Dry Basis)				
	Volatile at 1200°F	0.29	0.77	0.39	0.44
	MoO_3	11.50	11.70	11.30	11.30
	NiO	3.92	3.68	4.05	3.90
	SiO_2	2.24	0.52	1.32	3.16
	SO_4	0.57	1.04	1.39	0.67
	Na_2O	0.17	0.36	0.32	0.41
II	Physical Properties				
	Particle diameter (mm)	1.76	1.72	1.33	1.34
	Bulk density (g/cm^3)	0.65	0.69	0.82	0.67
	Side Crush Strength (N/mm)	10.73	7.67	11.20	8.25
	Surface Area (m^2/g)	177.0	118.0	210.0	101.0
	P. V. (H_2O) (ml/g)	0.52	0.38	0.43	0.42

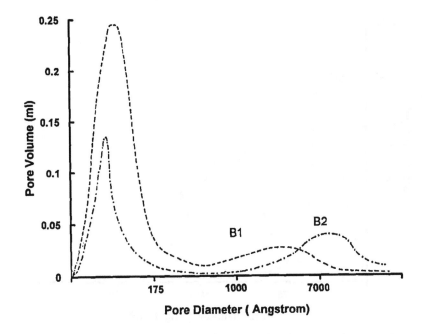

FIGURE 3. *Pore size distribution of the bimodal alumina extrudates.*

Both surface area and total pore volume of the prepared catalysts are relatively lower than typical commercial catalysts. Both values were initially low for the unimpregnated alumina extrudates. Impregnation with the active metals further reduced the surface area by 10-15%, apparently due to blockage of some of the micropores of the support. The pore size distributions of the bimodal and monomodal catalyst supports are shown in Figures 3 and 4, respectively.

Upon impregnation with the active metals, it was noted that the macropores of both bimodal catalysts were practically not affected. However, the mesopores were reduced both in terms of the their contribution to the total pore volume and their average diameter. Catalyst B1 had its mesopore peak at 75Å, lower by 10Å from the alumina support. This peak contributed around 85% of the total pore volume. On the other hand, the macropore peak was centered at 1800Å. For the B2 catalyst, the mean diameter for the mesopores was 65Å and the contribution of these pores to the total pore volume is ~ 60%.

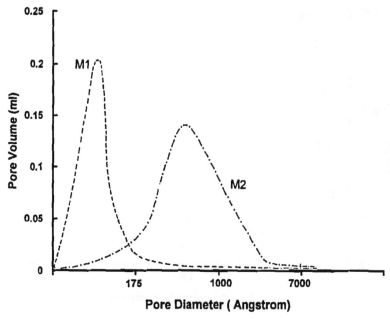

FIGURE 4. *Pore size distribution of the monomodal alumina extrudates.*

For the monomodal catalysts, the distributions of the pores of the impregnated samples were observed to shift toward lower average diameter. Catalyst M1 had a pore maxima of 80Å, and the pore diameter distribution ranging between 30-150Å. Similarly, catalyst M2 pore size distribution shifted to a narrow range between 150-350Å with a maxima around 230Å.

Activity of the Newly Developed Catalysts. The prepared catalysts were initially screened by assessing their performance toward hydrodesulfurization in a microactivity test using a feed consisting of gas oil mixed with 20% atmospheric residue. The results are summarized in Table 5. The results show that there is no clear correlation between the initial activity of the catalysts and their total pore volumes or surface areas.

The bimodal catalysts were observed to follow the same order of reactivity as that normally anticipated based on surface area and pore volume. The catalyst with larger surface area, B1, is considerably more active than B2.

TABLE 5. *Initial desulfurization activity of the prepared catalysts.*

Catalyst	Initial Desulfurization Activity (wt.%)	Surface Area (m^2/g)	Pore Volume (ml/g)
B1	60	177.0	0.52
B2	26	118.0	0.38
M1	40	210.0	0.43
M2	45	101.0	0.42

For the monomodal catalysts M1 and M2, the results of the microactivity hydrodesulfurization tests show an inverse relation with surface area. Thus, despite the lower surface area of M2, it was observed to be slightly more active than M1. This is presumably due to the larger pores of the former catalyst which do not get plugged upon the deposition of carbon. Furthermore, the test indicates that the bimodal catalyst, B1, is more active than both monomodal catalysts. However, further testing revealed that this higher activity is limited to the initial few hours of the catalyst life.

To assess the deacivation behavior of the catalysts, their performance was evaluated using a fixed bed pilot plant. The test is characterized by its severity with a feed consisting of around 80% vacuum bottoms, 20% heavy vacuum gas oil, a temperature of 400°C and pressure of 120 bar. The details of the feed characterisitcs are presented in Table 1. The test duration is 144 h, providing good assessment for both the activity and initial deactivation behavior of the catalysts.

The results of the bimodal catalyst performance in comparison with the monomodal catalyst M1 are shown in Fig. 5. Both bimodal catalysts were inferior to the monomodal catalyst, M1, throughout the duration of the run.

The catalyst with the larger macropores, B2, showed a lower rate of initial deactivation than B1. However, after around 72 hours on stream, both catalysts had similar activities till the end of the run. For the monomodal catalyst series, the overall performance of the catalyst M2, which has its pore distribution in the range 150-350Å and a pore maxima at 230Å, is higher than all other catalysts including the bimodal catalysts (Fig. 6). An interesting observation in the performance of this catalyst is the initial increase in its desulfurization activity during the initial 40 hours on stream. This phenomenon was confirmed by repeated tests and by the behavior of the catalyst in the microreactor test. This is attributed to the large mesopore diameter of this catalyst, which results in maintaining the pores of the catalysts open during the coking phase of the deactivation and possible redistribution of the active Ni/Mo phase during the initial stages of the reaction.

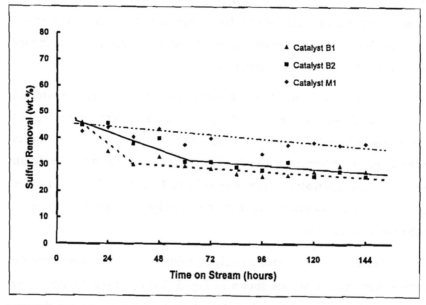

FIGURE 5. *Pilot plant test results for the bimodal catalysts vs the monomodal M1 catalyst.*

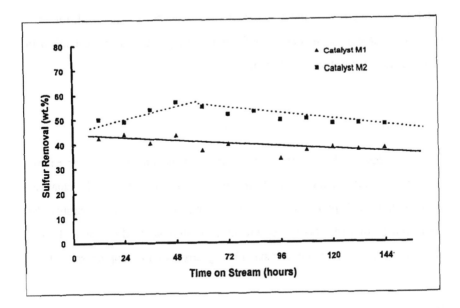

FIGURE 6. *Pilot plant test results for the monomodal catalysts.*

The above results demonstrate clearly the importance of pore size distribution in designing catalysts for residues hydroprocessing. It appears that despite the large pore volume of bimodal catalysts, the activity of these catalysts drops down significantly during the first few hours in stream. The cause of this drop seems to be due to the deposition of coke which causes the blockage of most of the micropores of the catalysts. As a result, the catalyst loses most of its active surface area. This is confirmed by a recent review by Absi-Halabi et al.(20), in which it was concluded that the initial deactivation of the catalysts is due to the deposition of large amounts of coke which blocks the smaller pores.

The activity of the catalysts seems to be optimum for catalysts with an average pore diameter of 150-350Å range. For the low pore diameter catalysts, the surface area appears to play a dominating role. For the larger pore diameter catalyst, M2, the results indicate that the pores do not get plugged, rather it seems that the deposited coke leads through some yet unknown mechanism to an increase in the initial activity of the catalyst.

Conclusion

The results of this study show clearly that monomodal catalysts having pore diameters in the range 150-350Å are more active than bimodal catalysts with similar total pore volume. Furthermore, the study demonstrates the effectiveness of hydrothermal treatment in designing catalysts with desired pore size distribution to meet the specific objectives of a hydroprocessing unit.

References

1.   Duncan, N. E. Oil & Gas J., May 4, 108(1992).

2.   Thakur, D.S., and M.G. Thomas. Applied Catalysis **45**, 197(1985).

3.   Speight J.G. "Upgrading heavy oils and residua". In Catalysis on the Energy Scene. Edited by S. Kallaguine and M. Mahay. Elsevier Scientific Publishing Company, pp. 515-527(1984).

4.   Takeuchi, C., Y. Fukui, M. Nakamura and Y. Shiroto. I & EC Process Design and Development **22**, 236(1983).

5.   Offenhauer, R.D.; J.A. Brennan and R.C. Miller. Industrial and Engineering Chemistry **49**, 1265(1957).

6.   Sie, S.T. "Catalyst deactivation by poisoning and pore plugging in petroleum processing". In Catalyst Deactivation. Edited by B. Delmon and G. F. Froment. Amsterdam: Elsevier Scientific Publishing Company, pp. 545-569(1980).

7.   Hennerup, P.N., and A.C. Jacobson. "A model for the deactivation of residue hydrodesulphurization catalysts". Presented at 185th National ACS meeting, Seatle, Washington, March 20-25(1983).

8.   Trimm, D.L. Applied Catalysis **5**, 263(1983).

9. Tamm, P.W., H.F. Harnsberger and A. G. Bridge. Industrial Engineering Chemistry Process Design Development 20, 262(1981).

10. Dautzenberg, F. M.; J. Van Klinken; K.M.A. Pronk; S.T. Sie; and J.B. Wijffels. "Catalyst deactivation through pore mouth plugging during residue desulphurization". ACS symposium series 65, 254(1978).

11. Pazos, J.M.; J.C. Gonzalex; and A. J. Salazar Gullien. Industrial Engineering Chemistry Process Design Development 22, 653(1983).

12. Shimura, M., Y. Shiroto and C. Takeuchi. Industrial Engineering Chemistry Fundamentals 25, 330(1986).

13. Thoulhoat, H., and J. Plumail. Studies in Surface Science and Catalysis 53, 463(1990).

14. Kobayashi, S., S. Kushiyama, R. Aizawa, Y. Koinuma, K. Inoue, Y. Shimizu, and K. Egi. Industrial Engineering Chemistry Research 26, 245(1987).

15. Fischer, R. H. and P. J. Angevine. Applied Catalysis 27, 275(1986).

16. Ahn, B. J. and J. M. Smith. AIChE J. 30, 739(1984).

17. Lahalieh, S, and M. Absi-Halabi. US Patent 4,820,766 (1989).

18. Absi-Halabi, M., A. Stanislaus and H. Al-Zaid, Stud. in Sur. Sc. Cat. 63,155(1991).

19. Absi-Halabi, M., A. Stanislaus and H. Al-Zaid. US Patent 5,217,940 (1993).

20. Absi-Halabi, M., A. Stanislaus and D. L. Trimm. Applied Catalysis 72, 193(1991).

12 A Study of Aluminophosphate Supported Ni-Mo Catalysts for Hydrocracking Bitumen

Kevin J. Smith, Department of Chemical Engineering,
University of British Columbia, Vancouver, Canada V6T 1Z4

Leszek Lewkowicz, Alberta Research Council,
Edmonton, Canada T0C 1E0

Mike C. Oballa and Andrzej Krzywicki,
Novacor Research and Technology Corporation,
2928-16 St. N.E., Calgary, Alberta, Canada T2E 7K7

INTRODUCTION

H-Oil and LC-Fining processes utilize a combination of thermal and catalytic
hydroprocessing reactions to achieve high yields of distillate in upgrading
bitumen or heavy oil residua. The processes are based on a well mixed
(ebullated bed) reactor from which deactivated catalyst is continuously
withdrawn and fresh catalyst is added to maintain yields. Catalyst activity and
lifetime are two key factors controlling the economics of these processes.

Catalyst deactivation occurs due to the deposition of coke and metals
on the catalyst surface. It is known that coke deposition is relatively rapid and
reaches a steady state value after some initial period whereas the metals
deposition continues to increase as more feedstock is processed (1). The metals
are associated with large asphaltene molecules present in the feed that cannot
penetrate small pore catalyst supports. Thus metals deposit on the periphery of
the catalyst particle and block the pore openings. The choice of catalyst is
usually a compromise between two extremes: small pore catalyst with low

metals capacity but higher activity that deactivates rapidly because of metals deposition and wide pore catalyst that has high metals deposition capacity but lower activity due to low surface area.

Recently, aluminophosphate materials with large pores (<10 nm - 1000 nm) and high surface areas (100 - 500 m^2/g) have been reported (2,3). The actual pore size distribution and surface area obtained depend on the Al/P ratio, preparation method and the calcination procedure. These materials are also thermally stable. The purpose of the present work was to determine if such materials, as a result of their pore size distribution and surface area, could decrease the rate of catalyst deactivation, increase catalyst activity and provide sufficient pore volume for high capacity of metals deposition during the upgrading of heavy oil residue.

EXPERIMENTAL

The Reaction Unit

All the catalyst performance data reported herein were obtained in continuous stirred-tank reactor systems. For the experimental runs of less than 12 hours duration, a 300 ml CSTR (Autoclave Engineers, Erie, PA) designed to operate at a maximum pressure of 28 MPa at a temperature of 500 °C, was used. The reactor was equipped with a stainless steel catalyst basket of 27 cc volume. Typical reaction conditions for this reactor were temperature 415-450 °C, hydrogen flow 1.0 SLPM and LHSV of $0.5h^{-1}$. The liquid holdup was kept constant at 130 cc and the catalyst loading was 11 grams.

A total of 13 runs were performed using this unit to identify the best hydrocracking reaction conditions, and to evaluate the aluminophosphate catalysts having a range of Al/P ratios. In all the runs the important process parameters such as temperature, pressure, liquid and gas flows were continuously monitored and recorded using precise instrumentation and control devices.

The apparatus used for longer time-on-stream studies (>12 hrs) was a Robinson-Mahoney reactor (Autoclave Engineers, Erie, PA) of 1 litre internal volume equipped with a stirrer and a specially designed catalyst basket. Sixty grams of catalyst was used for each run, the liquid hourly space velocity (LHSV) was $0.5 \, h^{-1}$, hydrogen flow rate was 1.0 SLPM, and pressure was 2300 psig. This reaction unit was described elsewhere (7), but suffice is to mention that it can be operated continuously with little operator intervention. It was therefore considered ideal for longer term runs in order to investigate the early deactivation profile of the catalysts.

Feedstock and Product Work-up

The Cold Lake atmospheric residuum feedstock were obtained from Alberta Research Council's Oil Sample Bank. The properties of the feedstock are presented in Table 1.

After reaction the liquid products were analyzed using a simulated distillation method (ASTM D2887) to obtain the yield of distillates (<524 °C fraction) and the pitch (+524 °C) conversion. In order to characterize the quality of the total reaction products the following analyses were performed: Conradson Carbon Residue (CCR), C, H, N and S elemental analysis and metals content (Ni+V) of the product. The following definitions are used herein to evaluate the efficiency of the hydrocracking reactions:

Pitch or 524 °C conversion = $[1-(524^+$g product$)/(524^+$g feed$)]$ x 100%

HDS conversion $\quad = \quad [1-(S, $g in product$)/(S, $g in feed$)]$ x 100%

CCR conversion $\quad = \quad [1-(CCR, $g product$)/(CCR, $g feed$)]$ x 100%

HDM conversion $\quad = \quad [1-(Ni+V, $wppm product$)/(Ni+V, $wppm feed$)]$ x 100%

Table 1: Properties of residua used in this study

Source	ESSO Cold Lake Vacuum Tower Bottoms
Viscosity [cP]	at °C: 80 : 6890.0 90 : 2875.0 100 : 1394.0 110 : 426.0
CCR [wt%]	16.6

Ultimate Analysis [wt%]:
  Carbon
  Hydrogen
  Nitrogen
  Sulphur
  Asphaltene (wt%)
	
	82.9
	10.4
	0.5
	5.0
	25.3*

Metals Content [wppm]:
  Vanadium
  Nickel
	
	233
	91

Simulated Distillation: IBP (°C) IBP - 524 °C, (wt%) Pitch 524 °C+ (wt%)	 350 Cut Temp [°C] . . . % Off 396 . . . . 4 425 . . . . 9 449 . . . . 14 474 . . . . 19 501 . . . . 24 525 . . . . 28.3 28 72

* Includes preasphaltenes

Catalyst Preparation and Pre-Treatment

The aluminophosphate catalyst supports of various Al/P ratios, were prepared by precipitation from aluminum nitrate/phosphoric acid solutions with ammonium hydroxide, following the procedure described by Kuo and Yang (3). A 0.5M solution of aluminum nitrate, prepared by dissolving $Al(NO_3)_3 \cdot 9H_3O$ in distilled water, was mixed with an appropriate amount of 85% H_3PO_4. This solution was added dropwise to a 14.5% NH_4OH solution that was stirred constantly. The pH was maintained at 8.0 throughout the precipitation by adding concentrated NH_4OH dropwise from a separate measuring funnel, as required. The gelatinous precipitate was filtered, washed and dried at 120 °C for 12 hours. The dry filter cake was subsequently ground to a powder, sized to < 30 mesh and calcined at 500 °C for 1 hour.

The Ni-Mo was dispersed on the aluminophosphate support by the incipient wetness technique from a single solution of $Ni(NO_3)$ and ammonium paramolybdate, $(NH_4)_6(Mo_7O_{24})$. The concentration of the solution was such that the Ni and Mo content of the final catalyst was approximately 2 wt% and 7 wt%, respectively. Following impregnation and drying at 120 °C for 24 hours, the catalysts were mixed with a bentonite clay (Volclay 325) and water to form a paste that was extruded through a 1/8" cylindrical die. The extrudates were dried at 120 °C for 2 hours and calcined at 500 °C for 1 hour prior to sizing to 9-20 mesh particles, to yield the catalysts tested in the present study. Before each test, the catalysts were presulfided by heating to 320 °C in a dimethyl disulphide/N_2 gas stream flowing over the hot catalyst for 45 minutes.

Following each experimental run the catalyst was transferred from the catalyst basket to an extraction thimble. It was extracted in toluene for 24 hours and dried in an oven, under a nitrogen blanket.

RESULTS AND DISCUSSION

Catalyst Properties

The physical and chemical properties of the catalysts studied in the present work are summarized in Table 2. Catalysts with four different Al/P ratios in the range 1 to 12 and a commercial Ni-Mo/Al$_2$O$_3$ catalyst were studied.

The surface areas of the aluminophosphate support materials decrease as the Al/P ratio decreases, whereas the pore volume was maximum at an Al/P ratio of 5.6 (catalyst TP 1732 F). Following dispersion of the Ni-Mo onto the supports and extrusion, both the surface areas and pore volumes were reduced. However, the trend in these measurements as a function of Al/P ratio remained the same. The data of Table 2 also show that the surface areas of the aluminophosphate supported catalysts are lower than the commercial catalyst, whereas the pore volume of the catalyst with Al/P = 5.6 is the same as that for the commercial catalyst. Analysis of the aluminophosphate by XRD showed that in all cases these materials were amorphous. The data show that as the Al/P ratio increased, the surface area increased and the average pore diameter (4 x pore volume/surface area) decreased, in accord with the results of Marcelin *et al* (4).

Effect of Al/P ratio on catalyst performance

The effect of Al/P ratio on catalyst performance was determined at 438 °C using Cold Lake feedstock. The reaction conditions were chosen to obtain approximately 65% pitch conversion, and the catalyst activities were determined within the first 12 hours time-on-stream. The results obtained for the 4 aluminophosphate based catalysts are presented in Table 3. In general, as the Al/P ratio increased the S, metals and CCR conversion increased. The pitch conversion data do not appear to follow a similar trend. However, considering

Table 2(a): Properties of precipitated aluminophosphate supported catalysts.

SAMPLE	SUPPORT AlPO$_4$			CATALYST Ni-Mo/AlPO$_4$		EXTRUDED CATALYST Ni-Mo/AlPO$_4$/Clay				
	Sg m²/g	Vg cc/g	d* nm	Sg m²/g	Vg cc/g	Sg m²/g	Vg cc/g	Al/P Ratio	Ni wt%	Mo wt%
TP 1733F	284	0.82	11.5	235	0.65	180	0.52	11.1	2.2	6.2
TP 1732F	190	1.11	23.4	152	0.82	125	0.67	5.56	2.2	6.5
TP 1731	108	1.01	37.4	--	--	76	0.66	1.96	2.0	6.0
TP 1734	49	0.57	46.5	28	0.24	31	0.34	1.14	2.2	7.4

* d = Average Pore Diameter
* XRD results indicate that all the supports are amorphous

Table 2(b): Properties of typical commercial NiMo/Al$_2$O$_3$ hydrocracking catalysts.

	A	B
Surface Area (Sg) [m²/g]	258	170
Pore Volume (Vg) [cm³/g]	0.67	0.58
Ni [wt%]	2.2	2.3
Mo [wt%]	6.16	10.0
Average Pore Diameter [nm]	104	136

Table 3: Effect of Al:P ratio on performance of Ni-Mo/AlPO$_4$ catalyst

Temperature................................... 438 °C

H$_2$............................ 2300 psig and 1.0 SLPM

LHSV............................. 0.5 h^{-1}

Catalyst Weight......... 11 g in 300 mL Robinson Reactor

Al/P RATIO*	CONVERSION			
	Pitch %	HDS %	HDM %	CCR %
11.1	68.3	77.5	86.4	56.8
5.56	64.2	66.3	87.9	50.5
1.96	65.1	58.6	85.5	48.1
1.14	68.1	57.8	81.2	46.0

* See Table 2

the error associated with the data, it is concluded that the pitch conversion is relatively independent of Al/P ratio.

In previous studies, P has been reported to increase the HDS and HDN activity of Ni-Mo/Al$_2$O$_3$ hydrotreating catalysts (5). Phosphorous addition also increased the S removal capability of dispersed Mo catalysts used in heavy oil hydrotreating (6). The present work has shown that for the aluminophosphate catalysts, the Al/P ratio is an important parameter that influences S, CCR and (Ni+V) removal activity of the catalyst. The data show that a high Al/P ratio is preferred and this corresponds to the catalyst with highest surface area and pore volume.

Deactivation by metals deposition will be particularly apparent over a longer time period than that investigated for the short-term experiments. The improvement due to pore structure was therefore investigated over +100 hrs using a second batch of two of the aluminophosphate catalysts (Al/P = 11.1 and 5.6, also designated TP1733F and TP1732F). These experiments were carried out in the 1L CSTR that had a higher catalyst-to-oil ratio than the 300 cc unit. In both cases, however, the reaction temperature was chosen so that pitch conversions were approximately 65%. The results are given in Figures 1-4. The figures show the dependence of HDM, CCR , HDS and pitch conversion on time-on-stream. In each graph, the corresponding performance of the base commercial catalyst is shown for purposes of comparison:

(1) Metals Conversion: TP1733F showed higher metals conversion for the interval studied. The two Al/P prepared catalysts out-performed the commercial catalyst.

(2) CCR Conversion: TP1733F showed a higher initial activity for CCR conversion than the other two catalysts. One would deduce from the figure that after 90 hours of run time, the activity of all the catalysts seemed to have levelled off at under 60%.

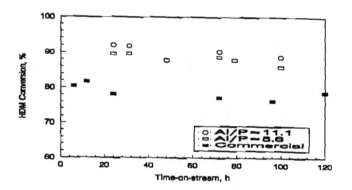

Figure 1:   HDM conversion as a function of time-on-stream at 403 °C, 15.6 MPa H$_2$ at 1.0 L/min STP, LHSV = 0.5 h^{-1} with 60g catalyst in 1L reactor.

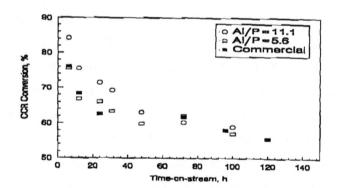

Figure 2:   CCR conversion as a function of time-on-stream at 403 °C, 15.6 MPa H$_2$ at 1.0 L/min STP, LHSV = 0.5 h^{-1} with 60g catalyst in 1L reactor.

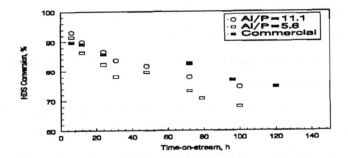

Figure 3: HDS conversion as a function of time-on-stream at 403 °C, 15.6 MPa
H_2 at 1.0 L/min STP, LHSV = 0.5 h^{-1} with 60g catalyst in 1L reactor.

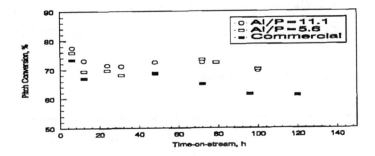

Figure 4: Pitch conversion as a function of time-on-stream at 403 °C, 15.6 MPa
H_2 at 1.0 L/min STP, LHSV = 0.5 h^{-1} with 60g catalyst in 1L reactor.

(3) <u>Sulphur Conversion</u>: Both TP1733F and the commercial catalysts showed higher sulfur conversion than TP1732F.

(4) <u>524 °C+ Conversion</u>: Both TP1733F and TP1732F out-performed the commercial catalyst.

The results show that TP1733F has a higher activity than TP1732F. Comparing the Ni-Mo/Al$_2$O$_3$ commercial catalyst with that of the aluminophosphate and noting that the commercial catalyst has higher surface area and presumbly an optimized Ni-Mo dispersion not present on the aluminophosphate catalyst, the activity data suggests a beneficial effect of the presence of the P introduced by the coprecipitation of the present study. With respect to the reactions, the results can be used to rank relative catalyst activity as follows:

Sulfur Conversion . . . . Commercial $\geq$ TP1733F > TP1732F

Metals Conversion . . . . . . TP1733F > TP1732F > Commercial

524 °C+ Conversion . . . . . .TP1733F$\geq$ TP1732F > Commercial

CCR Reduction . . . . .TP1733F$\geq$ Commercial > TP1732F

Table 4: Elemental analysis of fresh and used catalysts.

	C wt%	H wt%	N wppm	S wt%
1. Commercial Ni-Mo/Al$_2$O$_3$				
Fresh	0.15	0.63	38	0.74
Used	21.62	1.23	3130	6.2
2. Ni-Mo/AlPO$_4$ (Al/P = 11.1)				
Fresh	0.09	0.53	120	<0.1
Used	20.13	1.03	3179	7.87
3. Ni-Mo/AlPO$_4$ (Al/P = 5.56)				
Fresh	0.15	0.37	77	0.3
Used	17.24	0.85	2634	3.8

Table 5: Properties of fresh and used catalysts

PROPERTIES	COMMERCIAL Ni-Mo/Al$_2$O$_3$		Ni-Mo/AlPO$_4$ Al/P = 11.1		Ni-Mo/AlPO$_4$ Al/P = 5.56	
	FRESH	USED	FRESH	USED	FRESH	USED
Ni	2.13	2.64	1.68	3.22	1.70	2.04
Mo	5.8	3.61	4.28	2.44	4.49	3.05
Al	36.4	26.15	27.9	18.51	26.70	19.7
Fe			0.51	0.42	0.39	0.34
V		4.02		6.65		2.71
K	227 wppm	366 wppm	657 wppm	921 wppm	458 wppm	726 wppm
P	634 wppm	0.11	2.5	1.87	5.3	4.0
Mg	0.31	0.31		0.38	0.31	0.34
Ca	627 wppm	299 wppm	0.125	957 wppm	0.12	730 wppm
S	0.74	6.2	<0.1	7.87	0.3	3.8
Surface Area (Sg) (m^2/g)	248	57	181	33	140	73
Pore Volume (Vg) (cm^3/g)	0.602	0.225	0.718	0.256	1.093	0.486
Avg. Pore Diameter (nm) d=4Vg/Sg x 10^4	97	158	159	308	312	267
ΔSg (% Reduction)	77		82		48	
ΔVg (% Reduction)	63		64		56	

Explanations have appeared in the literature on the advantages of phosphorous in cobalt/nickel and molybdenum-type catalysts. Less coke formation is attributed to changes in acidity properties of the catalyst (8,9). Our results show the carbon content of both aluminophosphate-used catalysts as being less than that of the used commercial catalyst (Table 4).

Table 5 shows the properties of fresh and used catalysts. Both the commercial catalyst and the Ni-Mo/AlPO$_4$ with Al/P ratio of 5.56 (TP1732F) were run for 120 hours each. The Ni-Mo/AlPO$_4$ with Al/P ratio of 11.1 (TP1733F) was run for 220 hours. A comparison of the reduction in surface area and pore volume shows that TP1732F has the lowest reduction (less pore plugging). Catalyst TP1733F, even after almost double the run time of the commercial catalyst, has almost similar reduction. These results give an insight into the long term operability of the catalyst and we conclude that both AlPO$_4$-supported catalysts are better than the commercial catalyst at the reaction conditions studied and for a time-on-stream of less than 200 hrs.

CONCLUSION

The observed activity of a catalyst in a heterogeneous system depends on, among other factors, the surface area which is accessible to the reactants. For porous catalysts, it is not only the total surface area which governs the reaction, but also the average pore size and the pore size distribution. These properties of the catalyst become even more crucial when the catalyst is processing large organic molecules as those found in residuum. It is envisaged that the large organic molecules will be restricted from entering small pores and, therefore, only active sites found within large pores can contribute significantly to the reaction. The results of this work show that the aluminophosphate supported catalyst, because of its pore size, is superior to the commercial catalyst in terms of removal of the large metal-containing compounds (hydrodemetallation). Neither its desulfurization, nor its 524 °C+ or CCR conversion suffered as a result of this property. Furthermore, for the aluminophosphate catalysts, a high Al/P ratio is preferred since this corresponds to the catalysts with the greatest surface area and pore volume.

REFERENCES

1)   J. M. Oelderik, S. T. Sie and D. Bode, *Appl. Catal.*, 47, 1-24, (1987).

2)   T. T. P. Cheung, K. W. Wilcox, M. P. McDaniel and M. M. Johnson, *J. Catal.*, 102, 10-20, (1986).

3)   P. S. Kuo and B. L. Yang, *J. Catal.*, 117, 301-310, (1989).

4)   G. Marcelin, R. F. Vogel and H. E. Swift, *J. Catal.*, 83, 42-49, (1983).

5)   P. J. Mangnus and A. D. van Langeveld, V. H. J. de Beer, and J. A. Moulijn, *Appl. Catal.*, 68, 161-16, (1991).

6)   S. Kushiyama, R. Aizawa, S. Kobayashi, Y. Koinuma, I. Uemasu, and H. Ohushi, *Ind. Eng. Chem. Res.*, 30, 107-111, (1991).

7)   W. Wong, M. C. Oballa, *Continuous Flow Hydrocracker Design Conditions, Safety Features and Operating Procedures*, NHRC Internal Report #00546, May, 1990.

8)   A. Mordiez, M. M. Ramirez de Agudelo and F. Hemandez, *Appl. Catal.*, 41, 261, (1988).

9)   C. W. Fitz and H. F. Rase, *Ind. Eng. Chem. Prod. Res. Dev.*, 22, 40, (1983).

ACKNOWLEDGEMENTS

The experiments at the Alberta Research Council were performed by Mr. Blaine Doherty. The continuous flow hydrocracking experiments at Novacor were performed by Mr. Chi Wong. Their efforts are much appreciated and gratefully acknowledged.

We are grateful to Novacor, Alberta Research Council and EMR/CANMET management for providing the resources for this work, and to Dr. Paul Sears, the scientific officer at CANMET for useful discussions and review of both our bi-monthly reports and the final report and also for his cooperation.

13 Two-Stage Hydrotreating of a Bitumen-Derived Middle Distillate to Produce Diesel and Jet Fuels, and Kinetics of Aromatics Hydrogenation

Sok M. Yui
Research Center, Syncrude Canada Ltd.
10120 - 17 Street, Edmonton, Alberta T6P 1V8 Canada

Abstract

The middle distillate from a synthetic crude oil derived from Athabasca bitumen was further hydrotreated in a downflow pilot unit over a typical NiMo catalyst at 330° to 400°C, 7 to 11 MPa and 0.63 to 1.39 h^{-1} LHSV. Feed and liquid products were characterized for aromatics, cetane index (CI) and other diesel specification items. Aromatics were determined by a supercritical fluid chromatography method, while CI was determined using the correlation developed at Syncrude Canada Ltd. Also feed and selected products were distilled into a jet fuel cut (150°/260°C) by spinning band distillation for the determination of smoke point and other jet fuel specification items. A good relationship between aromatics content and CI was obtained. Kinetics of aromatics hydrogenation were investigated, employing a simple-first order reversible reaction model.

Introduction

Syncrude Canada Ltd. operates a surface mining oil sand plant at the Athabasca oil sand deposit in northern Alberta and produces synthetic crude oil (SCO) from the extracted bitumen. The virgin light gas oil (LGO) in the bitumen is separated in a distillation unit. The topped bitumen is cracked in two fluid cokers and an ebullated-bed hydrocracker into naphtha, LGO, and heavy gas oil (HGO). These streams are hydrotreated in separate reactors and then combined to form an SCO. A typical composition of SCO is 14 to 19 vol % naphtha (C_5/177°C), 44 to 48 vol % LGO (177°/343°C) and 33 to 39 vol % HGO (343°/520°C). Refiners use the LGO fraction as a diesel or jet fuel blending stock. LGO from synthetic crude is high in aromatics and does not by itself meet either the minimum cetane number (CN) specification (i.e., 40) for

235

diesel or minimum smoke point specification (i.e., 25 mm) for jet fuel. These are the main problems with bitumen-derived LGO that concern SCO customers and that need to be overcome.

In order to gain an understanding of the parameters affecting SCO quality during hydrotreating, we have carried out numerous experiments with commercial catalysts using various feedstocks: LGO and HGO produced at CANMET by hydrocracking Cold Lake heavy oil vacuum residue[1]; coker, hydrocracker and virgin LGOs[2]; coker and hydrocracker HGOs[3]; combined coker LGO and HGO[4], and coker naphtha[5]. Yui and Sanford[2] conducted hydrotreating experiments using LGOs from virgin, coked and hydrocracked Athabasca bitumen over typical commercial NiMo/Al$_2$O$_3$ catalysts. The feeds and products were distilled into diesel and jet fuel fractions and major product spec items were determined. Yui and Sanford[6,7] investigated kinetics of aromatics hydrogenation using a first order reversible reaction. The model was applied to published data on hydrogenation of aromatics determined by various methods[6] as well as to our own experimental data on ^{13}C NMR aromaticity over commercial NiMo catalysts with five feeds, i.e., hydrocracker LGO, virgin LGO, coker LGO, coker HGO, and combined coker LGO and HGO[7].

In the present study, further hydrotreating of the middle distillate (diesel fuel) fraction from the synthetic crude was conducted in a downflow pilot unit with a typical NiMo catalyst. Aromatics were determined by a supercritical fluid chromatography (SFC) method and CI by the correlation developed at Syncrude Canada Ltd.[2,8]. Kinetics of aromatics hydrogenation was investigated employing the same model previously developed[7]. The feed and selected products were distilled into jet fuel cuts (150°/260°C) by spinning band distillation to provide samples for the determination of smoke point, naphthalenes and other jet fuel specification items.

Experimental Section

The feed was the middle distillate fraction from SCO produced at the Syncrude plant. This fraction, which was hydrotreated product from the commercial hydrotreating units, was hydrotreated over a typical commercial NiMo/Al$_2$O$_3$ catalyst (3.8 wt % NiO, 20.6 wt % MoO$_3$, 162 m^2/g surface area, and 0.41 cm^3/g pore volume). The 120 mL (83.9 g) of catalyst charge was diluted with the same volume of 46 mesh silicon carbide.

The reactor (1.7-cm i.d. and 122-cm overall length) was heated by three independently-controlled electrical heaters. These heaters were controlled to give a uniform temperature throughout the reactor length, and a weighted number average was used as the isothermal temperature. The temperature was measured by a movable thermocouple on the outside skin. In a separate test using movable thermocouples at the reactor center and skin, it was shown that the radial temperature gradient was within 1°C. The catalysts were activated by sulfiding with paraffinic kerosene containing 1 wt % sulfur equivalent of 1-butanethiol. Steady-state activity was attained about 5 days after initial feed was charged and subsequently about 8 hours after changes in experimental conditions. Samples were collected when the reactor was in steady-state, practically once every working day.

Results

Table 1 summarizes general feed properties, including diesel fuel specification items.

Table 1 Properties of feed

Description	Unit	Properties	Type A Diesel Specifications
Density @20°C	g/mL	0.8673	
Sulfur	wppm	499	500 max[1]
Nitrogen	wppm	136	
Carbon	wt %	86.06	
Hydrogen	wt %	12.78	
SFC Aromatics	wt %	41.1	
Aniline Point	°C	47.8	
CN (Measured)	-	33.3	40 min
CN (Predicted)	-	34.1	
Kinematic Vis. @40°C	cSt	2.8	4.1 max
Flash Point	°C	75	40 min
Cloud Point	°C	-39	-34 max
RCR at 10% Bottom	mass %	0.10	0.15 max
Copper Corrosion	-	1b	1 max
Pour Point	°C	<-65	
D86 Distillation	°C		
IBP		187.0	
10%		215.5	
30%		242.2	
50%		263.1	
70%		286.0	
90%		313.1	315 max
FBP		333.7	

[1]As per the U.S. Environmental Protection Agency.

Yui

Table 2 summarizes hydrocarbon types of feed as obtained by mass spectrometry. Experiments were undertaken by varying the reactor temperature (330 to 400°C), pressure (7, 8.8, and 11 MPa), and LHSV (0.6 to 1.4 h^{-1}). LHSV (liquid hourly space velocity) is defined as volume of feed per hour per volume of catalyst loaded (120 mL) in the reactor. The system was operated once-through, and the hydrogen was of 100% purity. The pressure measured was the reactor total pressure, and the total pressure was considered to be the hydrogen partial pressure. The H$_2$/feed ratio was maintained constant (about 750 S m^3/m^3). Twenty-five total liquid product (TLP) samples were collected and characterized. Tables 3-1 and 3-2 give some results at typical operating conditions.

Table 2 Hydrocarbon types of feed by mass spectrometry (wt %)

Saturates	62.54		
Paraffins		17.63	
Cycloparaffins (Naphthenes) (C_{N0})	44.90		
Monocycloparaffins		23.22	
Dicycloparaffins			17.06
Polycycloparaffins		4.62	
Aromatics (C_{A0})	37.46		
Monoaromatics		34.40	
Alkylbenzenes			14.41
Benzocycloalkanes		13.78	
Benzodicycloalkanes		6.21	
Diaromatics		2.19	
Naphthalenes			2.04
Naphthocycloalkanes		0.15	
Fluorenes		0.00	
Aromatic Sulfur		0.88	
Benzothiophenes			0.85
Dibenzothiophenes		0.03	
Total	100.00		
Naphthenes (C_{N0})/Aromatics (C_{A0}) (= M)	1.20		

Table 3-1 Operating conditions and product properties - 1

RUN NUMBER		2	10	12	8	9	11
OPERATING CONDITIONS							
Temperature	°C	329.7	348.9	349.6	349.6	349.8	349.5
Pressure	MPa	8.8	8.8	7.0	8.8	8.8	11.0
LHSV	v/h/v	0.98	0.63	1.03	1.00	1.35	1.00
H_2/Feed	Sm^3/m^3	741	809	710	732	757	729
PRODUCT PROPERTIES							
Density @20°C	g/mL	0.8598	0.8572	0.8632	0.8609	0.8625	0.8582
Sulfur	wppm	15	5.1	7.5	38	13	6.1
Nitrogen	wppm	2.6	0.3	0.4	3.9	1.1	0.5
Carbon	wt %	87.65	87.43	86.78	87.58	86.18	85.79
Hydrogen	wt %	12.08	12.22	13.02	12.23	13.27	13.53
SFC Aromatics	wt %	33.2	25.7	35.3	30.6	33.7	26.8
Aniline Point	°C	53.6	58.2	52.0	55.0	53.0	57.7
CN (Predicted)	-	36.8	37.8	35.4	36.5	35.8	37.6
Kin. Vis. @40°C	cSt	2.8	2.7	2.8	2.8	2.8	2.8
Sim. Distillation	°C						
1%		149	142	148	146	148	144
5%		180	177	179	179	179	177
10%		195	192	195	195	196	193
30%		235	232	234	235	235	233
50%		266	262	265	264	265	264
70%		296	292	295	293	294	293
90%		331	328	330	329	330	329
95%		344	343	344	343	343	343
99%		363	363	365	363	363	363
IBP/177°C	wt %	4.4	5.1	4.5	4.6	4.4	4.5
177/343°C	wt %	90.2	90.1	90.1	90.5	90.5	90.1
343°C+	wt %	5.4	4.9	5.4	4.9	5.1	4.9

Table 3-2 Operating conditions and product properties - 2

RUN NUMBER		17	21	24	20	22	23
OPERATING CONDITIONS							
Temperature	°C	369.4	389.4	389.8	389.6	390.2	389.8
Pressure	MPa	8.8	8.8	7.0	8.8	8.8	11.0
LHSV	v/h/v	1.00	0.71	1.01	0.97	1.39	0.93
H_2/Feed	Sm^3/m^3	727	719	724	750	734	785
PRODUCT PROPERTIES							
Density @20°C	g/mL	0.8500	0.8427	0.8528	0.8463	0.8514	0.8427
Sulfur	wppm	3.5	167	29	136	104	83
Nitrogen	wppm	0.0	22	5.3	18.0	10	28
Carbon	wt %	85.72	84.66	86.60	85.22	86.11	87.44
Hydrogen	wt %	13.47	13.33	12.64	13.11	13.11	12.42
SFC Aromatics	wt %	24.4	25.1	35.1	27.3	30.4	20.0
Aniline Point	°C	58.5	57.5	51.0	56.7	53.5	61.8
CN (Predicted)	-	39.0	39.2	36.1	38.6	37.0	40.8
Kin. Vis. @40°C	cSt	-	-	-	-	-	-
Sim. Distillation	°C						
1%		102	75	86	80	96	82
5%		165	138	152	142	157	144
10%		184	170	179	173	180	174
30%		226	216	222	218	223	218
50%		258	249	254	252	255	251
70%		288	282	286	285	287	283
90%		327	326	326	330	326	325
95%		342	345	342	352	342	343
99%		362	408	368	429	367	382
IBP/177°C	wt %	7.7	12.1	9.4	11.1	8.9	10.9
177/343°C	wt %	87.7	82.4	85.9	82.3	86.4	84.2
343°C+	wt %	4.6	5.5	4.7	6.6	4.7	5.0

Discussion

Cetane Index

The data in Table 1 indicate that with the exception of CN the feed (diesel fraction from the Syncrude SCO) meets all product specifications, e.g., sulfur content, viscosity, flash point, cloud point, Ramsbottom carbon residue (RCR) and D86 distillation at 90%. The measured CN is 33.3 versus the specification minimum of 40. The predicted CNs or CIs in Tables 1 and 3 were obtained using the following correlation developed by Yui and Sanford[2,8]:

$$CI = -6979.4 + 7040.55*DE^2 - 13997.1*DE*\ln(DE)$$
$$- 7.91843*AP*\ln(DE) + 0.00771926*AP*MP$$
$$- 0.546587*AP*\ln(MP) - 0.00024134*MP^2 \qquad (1)$$

where DE is density (g/mL) at 20°C, AP is aniline point (°C) and MP is mid-boiling point (°C) determined by ASTM D2887 simulated distillation. Eq. (1) was developed expressly for synthetic distillates from northern Alberta bitumen using a total of 185 data points.

In an unpublished study which was totally separate from either the work of Yui and Sanford[2,8] or the work reported herein, CNs were measured for hydrotreated products obtained from hydrocracker, virgin and coker LGOs. Table 4 lists those numbers and compares them with predictions obtained by applying Eq. (1) and the ASTM D976-80 correlation to those previously unpublished data. The CNs predicted using Eq. (1) agree with engine measurements within the limits of reproducibility (±2.0 CN) of ASTM D613. This is quite remarkable given that the test engine and operator were different from those used in the development of Eq. (1).

Table 4 Measured and predicted cetane numbers of hydrotreated products

Feed Origin		Hydrocracker/Virgin LGO			Coker Gas Oil		
Cut Range	°C	243/290	290/343	343/399	243/290	290/343	343/399
CN (Measured)		33.9	38.7	40.4	30.8	33.8	32.7
CI by Eq. (1)		33.4	39.5	40.8	30.9	30.9	31.0
CI by ASTM D976		38.8	41.1	40.6	35.3	38.2	37.7
Density @15°C	g/mL	0.8740	0.8930	0.9158	0.8838	0.9104	0.9313
Density @20°C	g/mL	0.8710	0.8900	0.9129	0.8807	0.9074	0.9284
Aniline Point	°C	54.8	58.6	64.9	39.6	47.4	53.4
MBP by D2887	°C	264	325	365	268	318	359
MBP by D86	°C	262	303	353	259	313	360

Figure 1 illustrates CI vs reactor temperature at various pressures and LHSVs. The figure shows that product CI increases with:
- increasing hydrogen partial pressure
- decreasing space velocity
- increasing reactor temperature up to 380°C beyond which CI decreases.

Aromatics Content

In Table 1 the feed SFC aromatics content (41.1 wt %) is 3.6 wt % higher than mass spectrometry aromatics (37.5 wt %). Large portions of the saturates and aromatics consist of cycloparaffins (naphthenes) and mono-aromatics, respectively.

Figure 2 illustrates product aromatics content by SFC as a function of reactor temperature at various hydrogen partial pressures and LHSVs. The figure shows the well-known trend, i.e., the aromatics content of the product decreases with increasing reactor temperature up to a point, but then rises as the temperature is further increased, which is just the inverse of the trend for CI in Figure 1. In Figure 2 the solid lines indicate the equilibrium limits and the dashed lines indicate experimentally achievable levels. The calculated values were obtained from the simple first-order reversible reaction model[7]. Assuming that:
- the system is plug flow,
- the aromatics convert to naphthenes which undergo reverse reaction and follow first-order kinetics,
- the effect of side reactions is negligible

and using a power form of hydrogen partial pressure, then the aromatics hydrogenation may be represented as

$$- dC_A/dt = k_f p_{H_2}{}^\beta C_A - k_r C_N \tag{2}$$

where C_A and C_N are the percentage of aromatics and naphthenes, t is residence time, k_f and k_r are the forward and reverse rate constants, p_{H_2} is hydrogen partial pressure, and β is the power term. The mass balance is

$$C_{A0} + C_{N0} = C_A + C_N \tag{3}$$

where C_{A0} and C_{N0} are the percentages of aromatics and naphthenes in the feed. If we express the fractional degree of hydrogenation by $X_A = 1 - C_A/C_{A0}$ and let $C_{N0}/C_{A0} = M$, then Eq. (2) can be written as

$$dX_A/dt = (k_f p_{H_2}{}^\beta - k_r M) - (k_f p_{H_2}{}^\beta + k_r)X_A \tag{4}$$

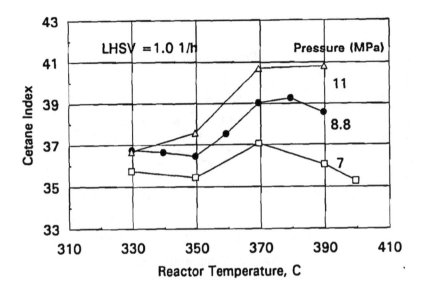

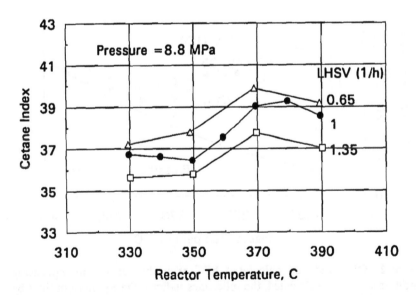

Figure 1 **Cetane index vs reactor temperature at various hydrogen partial pressure and LHSVs**

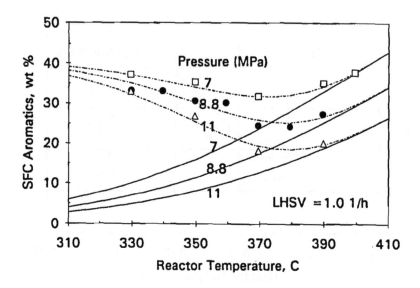

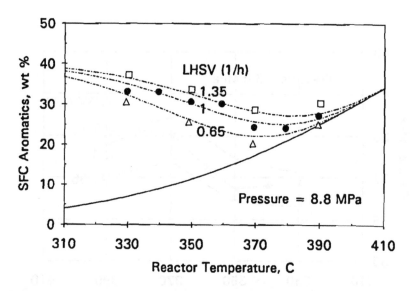

Figure 2 Observed and calculated SFC aromatics at various operating conditions. For the calculated, the solid lines indicate the equilibrium limit by Eq. (6) and the dashed lines indicate achievable limit by Eq. 5 where X_A and X_{Ae} are $1 - C_A/C_{A0}$.

The solution to Eq. 4 is

$$X_A = [(k_f p_{H_2}^{\beta} - k_r M)/(k_f p_{H_2}^{\beta} + k_r)]$$
$$[1 - \exp\{-(k_f p_{H_2}^{\beta} + k_r)/LHSV\}] \qquad (5)$$

where 1/LHSV is assumed to be the residence time t. The fractional degree of hydrogenation at equilibrium, X_{Ae}, is obtained by putting the left hand side of Eq. (4) equal to zero:

$$X_{Ae} = (k_f p_{H_2}^{\beta} - k_r M)/(k_f p_{H_2}^{\beta} + k_r) \qquad (6)$$

The rate constants k_f and k_r are assumed to follow an Arrhenius type equation:

$$k_f = k_{f0} \exp(-E_f/RT) \qquad (7)$$

$$k_r = k_{r0} \exp(-E_r/RT) \qquad (8)$$

where k_{f0} and k_{r0} are frequency factors, E_f and E_r are activation energies, R is the gas constant and T is the absolute temperature.

In the previous study on the kinetics of the hydrogenation of ^{13}C NMR aromatic carbon[7], we assumed from the stoichiometry of aromatic carbon reduction that the power term β for hydrogen partial pressure was 0.5 and that the heat of reaction H_r (= E_r - E_f) was 32.5 kJ/mol of aromatic carbon.

In the present study, kinetic parameters k_{f0}, k_{r0}, E_f and E_r and the power term β for hydrogen partial pressure in Eqs. (5), (7) and (8) were simultaneously determined employing a commercial nonlinear regression program[9]. M is 1.20 as shown in Table 2. The results are summarized as part of Table 5. The power term β and heat of reaction H_r are obtained as 1.73 and 84.65 kJ/mol of aromatics, respectively. In Figure 2 it is noted that at 370° to 380°C a higher degree of hydrogenation can be achieved at higher pressure and lower space velocity. This phenomenon is consistent with our previous findings[6,7].

Figure 3 illustrates CI as a function of SFC aromatics. The following inverse relationship is obtained:

$$CI = 45.8 - 0.282 * SFC \ Aromatics \ (wt \ \%) \qquad (9)$$

Figure 3 shows that 40 CI can be obtained by reducing the product aromatics to 20 wt % which is achievable at 370°C, 1 h⁻¹ LHSV and 11 MPa as shown in Figure 1.

Table 5 Kinetic parameters of aromatics hydrogenation in Eqs. (5) to (8), and reduction of density and mid-boiling point in Eq. (11)

	Hydrogenation of Aromatics	Reduction of Density	Reduction of Dist. @50%
No. of Observations	24	24	24
r^2	-	0.8712	0.9403
$M\ (=C_{N0}/C_{A0})$	1.20	-	-
α for LHSV	1.00	0.79	0.69
β for p_{H_2}	1.73	1.29	1.57
E_f (kJ/mol)	125.3	-	-
E_r (kJ/mol)	210.0	-	-
$k_{f0}\ (h^{-1}\,MPa^{-\beta})$	$2.979*10^8$	-	-
$k_{r0}\ (h^{-1})$	$2.432*10^{16}$	-	-
E/R (K)	-	8.416	17.107
$\ln k_0$	-	6.098	19.601

Aniline Point

Figure 4 illustrates SFC aromatics as a function of aniline point. The following relationship is obtained:

$$\text{SFC Aromatics (wt \%)} = 116.7 - 1.574 * \text{Aniline Point (°C)} \qquad (10)$$

According to our previous study[8], the linear correlation is valid only within the same oil group.

Density and Mid-boiling Point

Figures 5 and 6 illustrate density and mid-boiling point as functions of various operating conditions. The points are lab data and the lines are predicted values. The predicted values were obtained by the following first order and Arrhenius-type kinetic equation, including the power terms α for LHSV and β for p_{H_2}:

$$\ln(DE_f/DE_p) \text{ or } \ln(MP_f/MP_p) = k_0 \exp(-E/RT)p_{H_2}{}^{\beta}/LHSV^{\alpha} \qquad (11)$$

where subscripts f and p are feed and product, k_0 is frequency factor, and E is activation energy. Table 5 summarizes kinetic parameters determined by a regression analysis.

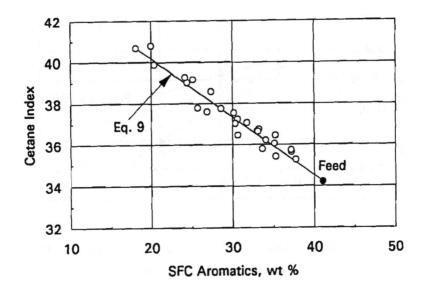

Figure 3 SFC aromatics content vs cetane index

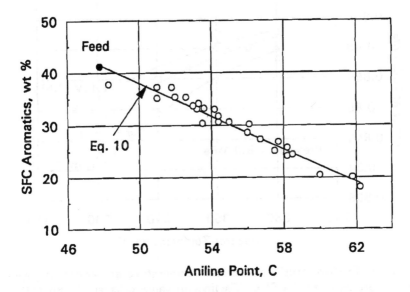

Figure 4 Aniline point vs SFC aromatics content

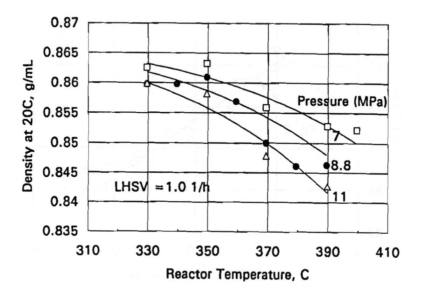

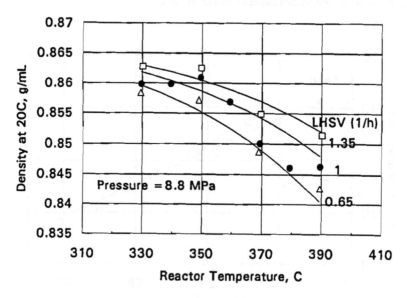

Figure 5 Product density vs reactor temperature at various hydrogen partial pressures and LHSVs. The lines are values calculated by Eq. (11).

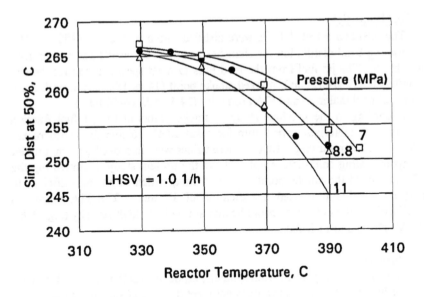

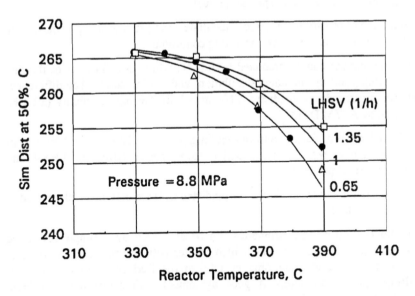

Figure 6 Product mid-boiling point vs reactor temperature at various hydrogen partial pressures and LHSVs. The lines are values calculated by Eq. (11).

Jet Fuel Properties

The feed and selected TLPs were distilled into a jet fuel cut (150°/260°C) by spinning band distillation for characterization. The results are summarized in Table 6. The jet fuel from the feed fails to meet specifications for density (0.845 vs 0.839 g/mL max), relaxed smoke point (14 vs 18mm min with 3 vol % max of naphthalenes) and FIA aromatics (34.4 vs 22 vol % max).

According to the previous study[2], smoke points of the jet fuel fraction from the raw materials were 9 mm for coker LGO, 16 mm for hydrocracker LGO, and 17 mm for virgin LGO. These values were improved by 3 mm under the hydrotreating conditions employed. In the present study hydrotreating the SCO improved the smoke point by 2 to 5 mm, which might be sufficient to meet the minimum relaxed specification of 18 mm. The FIA aromatics specification (20 vol % min) could be met at relatively high severity (e.g., 1 h^{-1} LHSV, 370°C and 11 MPa).

Conclusions

• Middle distillate separated from synthetic crude oil, itself a blend of bitumen-derived hydrotreated distillates, does not meet the 40 CN spec for diesel or the 18 mm relaxed smoke point spec for jet fuel.

• Further hydrotreating the middle distillate does produce on-spec diesel and jet fuels but at relatively severe operating conditions.

• Cetane number can be predicted by Eq. (1) which is a function of density, mid-boiling point by simulated distillation, and aniline point. The differences between measured and predicted CNs are within the reproducibility (± 2 CN) specified by ASTM D613.

• The reduction of density and mid-boiling point can be expressed by a first-order Arrhenius type kinetic equation using power terms for LHSV and hydrogen partial pressure.

• Aniline point and cetane index (or predicted CN) are both inversely related to aromatics content.

• Hydrogenation of SFC aromatics can be expressed by a simple reversible first-order kinetic equation which incorporates both forward and reverse rate constants, the naphthenes/aromatics ratio of the feed and a power term for hydrogen partial pressure. A good agreement with actual and calculated aromatics was observed.

Table 6 Properties of jet fuel fraction (150°/260°C)

| Description | Unit | Feed | Hydrotreated Products | | | Jet A-1 |
			12	18	16	Specifications
RUN NUMBER			12	18	16	
OPERATING CONDITIONS						
Temperature	°C	-	349.6	369.4	369.4	
Pressure	MPa	-	7.0	7.0	11.0	
LHSV	v/h/v	-	1.03	1.01	0.98	
PRODUCT PROPERTIES						
Density @15°C	g/mL	0.8447	0.8399	0.8386	0.8339	0.774-0.839
Sulfur	wppm	60	20	6	<5	2000 max
Nitrogen	wppm	20	3	2	<1	
Carbon	wt %	86.12	86.32	85.53	86.46	
Hydrogen	wt %	12.85	13.02	12.93	13.48	
Smoke Point	mm	14.1	16.4	16.4	19.2	25 min[1]
Naphthalenes	vol %	0.57	0.23	0.28	0.11	3 max[1]
FIA						
Saturates	vol %	63.4	71.3	72.7	83.4	
Oleffins	vol %	2.2	1.7	1.9	1.6	
Aromatics	vol %	34.4	27.0	25.4	15.0	22 max
SFC Aromatics	wt %	34.5	28.7	27.8	15.9	
Vis. @-20°C	cSt	6.57	6.06	5.59	6.19	8 max
Flash Point	°C	63.5	57.0	47.5	48.5	33 min
Freezing Point	°C	<-65	<-65	<-65	<-65	-47 max
Acidity	mmKOH/g	0.02	0.03	0.01	0.00	0.1 max
D86 Distillation	°C					
IBP			176.5	170.0	165.1	161.5	
10%			199.1	196.0	194.0	192.1	204 max
30%			213.4	209.7	209.2	207.6	
50%			222.7	218.1	218.2	217.8	Report
70%			231.1	226.4	226.3	227.9	
90%			243.4	237.7	237.1	241.1	Report
FBP			262.1	254.4	252.8	258.0	300 max

[1] Smoke point of 20 min plus 3 vol % max of naphthalenes.
Smoke point of 18 min plus 3 vol % max of naphthalenes is permitted provided purchaser is notified within 30 days of such shipment.

Acknowledgements

The author wishes to thank the following: Mr. Dave Famulak for operation of the Syncrude pilot plant, the analytical technologists at Syncrude Research and at the Alberta Research Council for analysis of samples, Dr. John Stone at Syncrude Research for review of this manuscript and Syncrude Canada Ltd. for permission to publish.

Literature Cited

1. Sanford, E.C. and S.M. Yui, "Hydrotreating Characterization of Coked and of Hydrocracked Gas Oils from Alberta Bitminous Oils with Commercial Ni-Mo Catalysts, and prediction of Some Product Properties,"Studies in Surface Science and Catalysis, 19, 585-592, Elsevier Sci. Pub. B.V., Proceedings, 9th Canadian Symp. on Catalysis, Quebec, P.Q., Canada, Sept. 30 - Oct. 3, 1984.

2. Yui, S.M. and E.C. Sanford, "Diesel and Jet Fuel Production from Athabasca Bitumen, and Cetane Number Correlation," Proceedings, 4th UNITAR/UNDP International Conference on Heavy Crude and Tar Sands, Edmonton, Alberta, Canada, Paper No. 13, 5, Aug. 7-12, 1988.

3. Yui, S.M. and E.C. Sanford, "Mild hydrocracking of Bitumen-Derived Coker and Hydrocracker Heavy Gas Oils: Kinetics, Product Yields, and Product Properties," Ind. Eng. Chem. Res. 28, 1278-1284 (1989).

4. Yui, S.M., "Hydrotreating of Bitumen-Derived Coker Gas Oil: Kinetics of Hydrodesulfurization, Hydrodenitrogenation and Mild Hydrocracking, and Correlations to Predict Product Yields and Properties," AOSTRA J. Res. 5, 211-224 (1989).

5. Yui, S. and E. Chan, "Hydrogenation of Coker Naphtha with NiMo Catalysts," Studies in Surface Science and Catalysis, 73, 59-66, Elsevier Sci. Pub. B.V., Proceedings, 12th Cannadian Symposium on Catalysis, Banff, Alberta, Canada, May 25-28, 1992.

6. Yui, S.M. and E.C. Sanford, "Kinetics of Aromatics Hydrogenation and Prediction of Cetane Number of Synthetic Distillates," Proc., API Refining Department, 50th Midyear Meeting, Kansas City, Mo., May 13-16, 1985.

7. Yui, S.M. and E.C. Sanford, "Kinetics of Aromatics Hydrogenation of Bitumen-Derived Gas Oils," Can. J. Chem. Eng. 69, 1087-1095 (1991).

8. Yui, S.M. and E.C. Sanford, "Predicting Cetane Number and ^{13}C NMR Aromaticity of Bitumen-Derived Middle Distillates from Density, Aniline Point, and Mid-Boiling Point," AOSTRA J. Res. 7, 47-53 (1991).

9. SAS User's Guide: Statistics, Version 5. SAS Institute Inc., Cary, NC (1985).

14 Aromatics Saturation Over Hydrotreating Catalysts: Reactivity and Susceptibility to Poisons

Peter Kokayeff
Unocal Science and Technology Division
376 South Valencia Avenue
Brea, California 92621

Abstract

Aromatics saturation over hydrotreating catalysts has been investigated by processing a synthetic feedstock to which naphthalene, biphenyl, tetralin, and cyclohexylbenzene have been added as model aromatic compounds.

The saturation reactions were successfully correlated with a kinetic model of consecutive first order reactions. The rates of saturation of the di-aromatics, naphthalene and biphenyl, were approximately 5 - 40 times faster than the rate of saturation of the mono-aromatics, tetralin and cyclohexylbenzene.

The nature of the di-aromatic is a significant determinant of its reactivity for saturation. The rate of saturation of naphthalene is 10 - 20 times faster than the rate of saturation of biphenyl. The two mono-aromatics, tetralin and cyclohexylbenzene, exhibited nearly identical reactivities for saturation.

An investigation of the effect of organo-nitrogen compounds on saturation activity revealed a very severe poisoning and attenuation of saturation activity. Since activity was recoverable upon the removal of the poisoning agent the chemical effect was conjectured to be due to adsorption of the poison on the active sites of the catalyst. The adsorption constants were determined for three model nitrogen compounds - quinoline, indole, and tert-butylamine. The poisoning action of an organo-nitrogen compound was found to be dependent on both basicity and chemical structure.

Introduction

The refining industry is focusing on meeting the new standards set for hydrocarbon fuels by the amendments to the Clean Air Act passed by Congress in October 1990. For diesel fuels the new specifications call for a maximum sulfur level of 0.05 wt % and an indirect limit on aromatics content via a specification of a minimum cetane index of 40. In addtition to these specifications refiners in California may have to meet an additional limit on aromatics content of a maximum of 10 vol %.

Although new specifications for diesel fuels relating to sulfur and aromatics have been anticipated for some time (1,2,3) the passage of the amendments to the Clean Air Act have resulted in an increased level of activity in this area as refiners re-examine the impact of this legislation on the economics of the industry (4,5,6). The options available to the refiner for achieving these limits have been described in a number of recent papers dealing with a detailed examination of the process chemistry of aromatics saturation and process conditions required to effect aromatics saturation and sulfur reduction (7,8,9,10) over hydrotreating catalysts.

Research efforts at Unocal Science and Technology Division have focused on the investigation of catalysts and processes that allow for a cost effective way to meet the new diesel fuel specifications. Pilot plant and bench scale investigations were carried out with both model compounds and commercial diesel blendstocks. One of the results of these investigations has been the observation of a very severe poisoning of the saturation activity of hydrotreating catalysts by nitrogen compounds (11). Although the attenuation of saturation activity of hydrotreating catalysts by nitrogen compounds has been reported in the literature (9), the severity of the poisoning action has not been fully appreciated.

Experimental

A series of experiments were performed with a synthetic feedstock consisting of 10 mol % Aromatic in Ethylflo 162.

Ethylflo 162 is a fully saturated poly-alpha olefin with a carbon number in the C_{15}-C_{30} range with an average molecular weight of 256, giving it an average carbon number of C_{20}. It is a clear, colorless, paraffinic hydrocarbon with a specific gravity of 0.797 and boiling at approximately 600 degrees F. Since these properties provided for a reasonable approximation of a paraffinic solvent for aromatic components which was otherwise pure and essentially inert it was chosen as the vehicle for carrying the aromatic components in these experiments.

The aromatics used included Naphthalene, Biphenyl, Tetralin, and Cyclohexylbenzene.

The catalyst used in all experiments was a commercially available Ni/Mo hydrotreating catalyst that had previously been in diesel hydrotreating service in pilot plant investigations. The catalyst was toluene washed and dried in air prior to use.

A simplified schematic of the reactor system used in these investigations is depicted in Fig. 1. The hydrocarbon feed, contained in a feed vessel, was delivered to the reactor by a Milton Roy mini-pump. Hydrogen was metered by a Brooks mass flow controller, mixed with the hydrocarbon, and the combined feed mixture entered a theromostatted reactor vessel. Exiting the reactor the products entered a vapor/liquid separator. Hydrogen, exiting the separator, was routed to a back-pressure regulator and then to the system vent. Liquid product was passed through a motor valve and then to a product collection vessel.

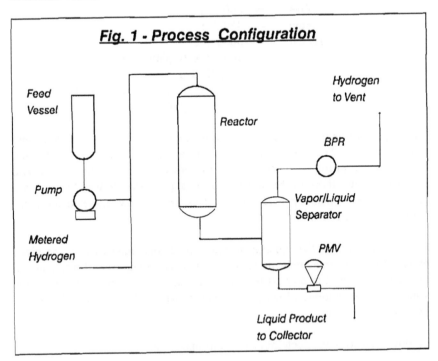

The following process conditions were employed in the saturation experiments:

LHSV	=	1.5/hr
Pres	=	600 psig
Temp	=	380 - 580 deg F
H_2/Oil	=	5000 SCF/B
Feed	=	10 mol % Aromatic in Ethylflo 162 with 1000 ppmw Sulfur as DMDS

The DMDS (dimethyl disulfide) was added to maintain the catalyst in a sulfided state.

A final experiment was conducted to investigate the effects of the presence of nitrogen compounds on catalytic activity. The feed was again changed to one containing naphthalene and doped to various nitrogen concentrations with organo-nitrogen compounds. For the investigations of the effects of nitrogen on saturation activity the following process conditions were used:

LHSV	=	1.5/hr
Pres	=	600 psig
Temp	=	450 deg F
H_2/Oil	=	5000 SCF/B
Feed	=	10 mol % Naphthalene Ethylflo 162 with 1000 ppmw Sulfur as DMDS and 0 - 50 ppmw Nitrogen

The organo-nitrogen compounds that were used were quinoline, indole, and tert-butylamine. These were used in concentrations of 1, 2, 4, 8, 15, 25, and 50 ppmw as nitrogen starting with the lowest concentration level and increasing sequentially to the highest. The catalyst was allowed to process a given nitrogen containing feed for two days prior to switching to the next, higher, nitrogen content.

At the conclusion of the experiment with a feed containing each of the given organo-nitrogen compounds the feed was switched to one free of nitrogen and the catalyst allowed to process this nitrogen-free feed for up to 15 days at the same process conditions allowing for a determination of the extent of recovery of saturation activity.

Product aromatics were determined by chromatography.

Finally, an experiment was conducted with a hydrotreated diesel feedstock at the following process conditions:

LHSV	=	1.5/hr
Pres	=	1000 psig
Temp	=	680 deg F
H_2/Oil	=	5000 SCF/B
Feed	=	Hydrotreated diesel feed with 1000 ppmw Sulfur as DMDS and 0 - 400 ppmw Nitrogen as Quinoline

The experiment was conducted with the clean feed doped to 1000 ppmw sulfur as DMDS (to keep the catalyst in the sulfided state) and with the same feed doped to nitrogen concentrations of 10, 25, 50, 100, 200, and 400 ppmw with quinoline. The catalyst was allowed to process a given nitrogen containing feed for two days prior to switching to one with the next highest nitrogen content. The aromatics content of the product was analyzed by FIA.

Discussion

Experiments with Nitrogen Free Feeds

First Order Consecutive Reaction Model - Fit to Experimental Data

A kinetic model consisting of two consecutive, irreversible, first order reactions, Fig. 2, was fit to the data from these experiments. The reverse reaction can be safely neglected since, at the temperatures of these experiments, an examination of the equilibrium position reveals that decalin and dicyclohexyl are the predominantly favored products.

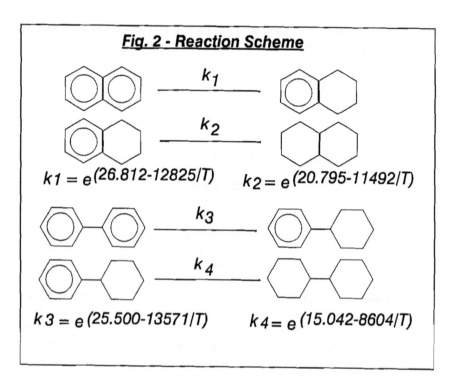

Fig. 2 - Reaction Scheme

$k1 = e^{(26.812-12825/T)}$ $k2 = e^{(20.795-11492/T)}$

$k3 = e^{(25.500-13571/T)}$ $k4 = e^{(15.042-8604/T)}$

The two rate constants were determined from the two separate experiments, i.e. the rate constant k_1 and k_3 were determined from the experiments with Ethylflo/Napthalene(Biphenyl) using only the data for the disappearance of the di-aromatic, while the rate constants k_2 and k_4 were determined from the experiments with Ethylflo/Tetralin(Cyclohexylbenzene) and were calculated from the disappearance of the mono-aromatics.

This method was used with the specific purpose of determining whether the postulated reaction scheme of consecutive reactions is valid. If the two sets of rate constants are determined from two independent experiments then one can assess the validity of the kinetic model by using both rate constants to generate curves for the yields of all components in the experiments with the di-aromatics. If this procedure results in a good fit to the data then one can be confident in the validity of the reaction scheme chosen, otherwise some other reaction network may need to be postulated. Clearly, sufficient data exists to determine both rate constants from the experiments with the di-aromatics (Ethylflo/Naphthalene(Biphenyl)) alone, but that procedure would not provide

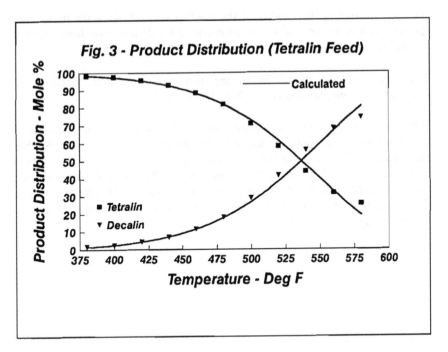

Fig. 3 - Product Distribution (Tetralin Feed)

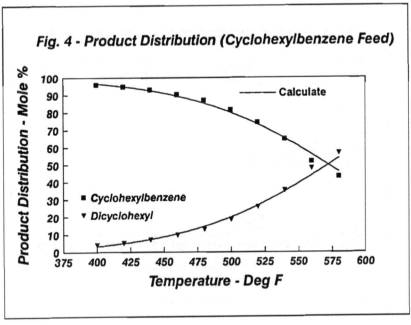

Fig. 4 - Product Distribution (Cyclohexylbenzene Feed)

for a verification of the validity of the reaction network since the parameters (pre-exponential factors and activation energies) would then have been adjusted to force a fit to the data.

The results of the model fit to the two mono-aromatics, tetralin and cyclohexylbenzene, are depicted in Fig. 3 and Fig. 4.

While it was satisfying to find a reasonably good fit to the data for the cases of the mono-aromatics (tetralin and cyclohexylbenzene), it is not altogether unexpected. The rate constants k_2 and k_4, determined from from these experiments, were then used with the rate constants k_1 and k_3, determined from the experiments with the di-aromatics (naphthalene and biphenyl) to calculate the entire product spectrum obtained with the di-aromatics, naphthalene and biphenyl, Fig. 5 and Fig. 6.

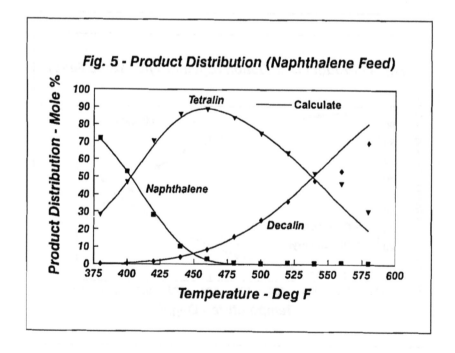

Fig. 5 - Product Distribution (Naphthalene Feed)

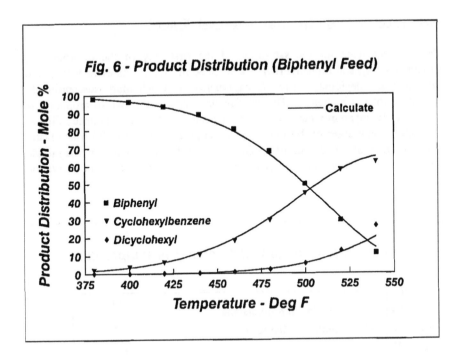

Fig. 6 - Product Distribution (Biphenyl Feed)

Although there is some breakdown at the higher temperatures, the overall fit is exceptionally good for all components over most of the temperature range.

The activation energies for the saturation reactions of the first rings of the di-aromatics were very similar with E_1 = 25,480 cal/gmol for the saturation of the first ring of naphthalene and E_3 = 26,970 cal/gmol for the saturation of the first ring of biphenyl. The activation energies for the saturation of the remaining rings on the mono-aromatic were considerably lower, E_2 = 22,840 cal/gmol for the saturation of tetralin and E_4 = 17,096 cal/gmol for the saturation of cyclohexylbenzene.

Additionally, the pre-exponential factors were also quite different. The pre-exponential factors for the saturation of the first ring of the di-aromatics were quite similar and much larger than the pre-exponential factors for the saturation of the remaining ring on the mono-aromatics.

The relative rates of saturation of the different types of aromatics may be assessed from the relative magnitudes of the rate constants for each reaction.

Saturation of the First Ring - Naphthalene and Biphenyl

The rate constants for the saturation of the first ring of the di-aromatics, naphthalene (k_1) and biphenyl (k_3), are depicted in Fig. 7. As evident from the rate constants depicted in Fig. 7, naphthalene is much more susceptible to saturation than is biphenyl. The relative reactivity of the two di-aromatics is more readily apparent when the ratio of the rate constants is examined, Fig. 8. The rate of saturation of the first ring of naphthalene is 10 to 20 times faster than the rate of saturation of biphenyl, with the difference in magnitude decreasing with increasing temperature due to the differences in activation energies.

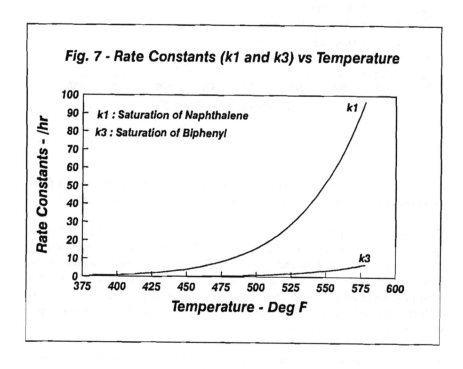

Fig. 7 - Rate Constants (k1 and k3) vs Temperature

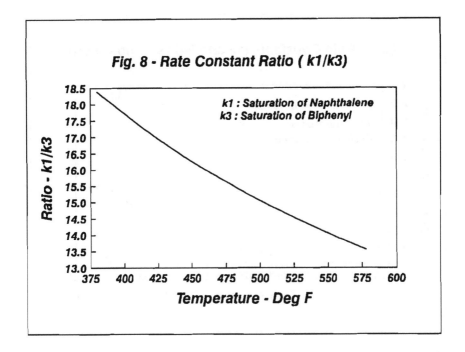

Fig. 8 - Rate Constant Ratio (k1/k3)

k1 : Saturation of Naphthalene
k3 : Saturation of Biphenyl

Saturation of the Second Ring - Tetralin and Cyclohexylbenzene

The rate constants for the saturation of the second ring of the resultant mono-aromatics, Fig. 9, are not nearly as different in magnitude as those for the saturation of the first ring of the di-aromatics. Relative reactivities of the two mono-aromatics, tetralin and cyclohexylbenzene, may be assessed by examining the ratio of the two rate constants k_2 (tetralin) and k_4 (cyclohexylbenzene), Fig. 10. The rate constant for the saturation of tetralin is only 0.6 - 2.0 times as large as that for the saturation of cyclohexylbenzene as opposed to a ratio of 10 - 20 which was observed for the case of the parent di-aromatics.

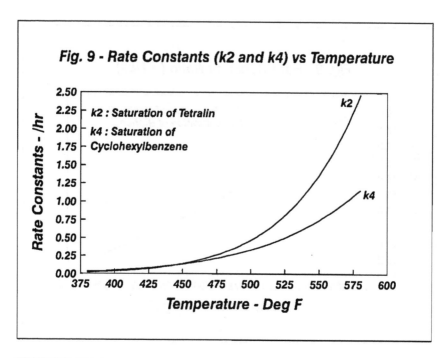

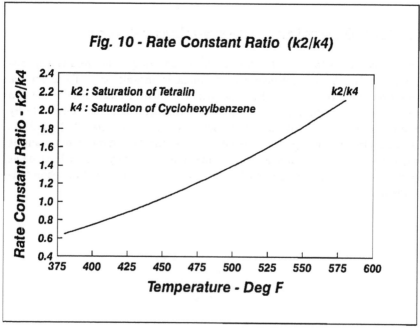

Saturation of Di- vs Mono-Aromatics

Finally, an examination of the ratios of the rate constants for the saturation of the di-aromatic and the corresponding mono-aromatic (Fig. 11) reveals that, whereas naphthalene undegoes saturation much faster than tetralin, the rate of saturation of biphenyl, although certainly greater than that of cyclohexylbenzene, is not an order of magnitude greater, as observed for the case of naphthalene/tetralin.

The rate of saturation of naphthalene (specifically of the first ring) is 20 to 40 times faster (when expressed as a ratio of the rate constants) as is the rate of saturation of tetralin, while the rate of saturation of biphenyl is only 1 - 8 times faster than the rate of saturation of cyclohexylbenzene. In both cases the rate of saturation of the first ring increases much faster with temperature than the rate of saturation of the second ring.

These observations lead to the conclusion that the reactivities of di-aromatics may vary over a much wider range than those of mono-aromatics. Additionally, the relative reactivities of a mono-aromatic and its parent di-aromatic may also vary over a wide range thus rendering any generalized statements as to relative reactivities of questionable accuracy.

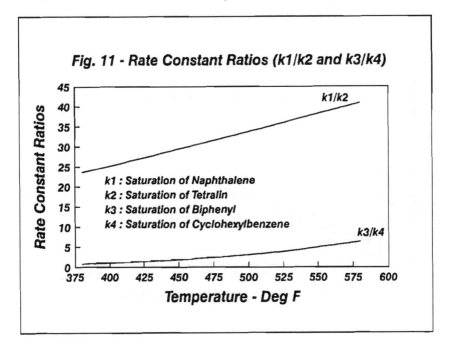

Fig. 11 - Rate Constant Ratios (k1/k2 and k3/k4)

Saturation of Naphthalene - Nitrogen Poisoning

Since industrial feedstocks contain nitrogen as well as sulfur, organo-nitrogen compounds were introduced into the feed. Three organo-nitrogen compounds were used: quinoline, indole, and tert-butylamine. These choices were motivated by the aim of investigating the effects of basic nitrogen (quinoline : pKa = 4.9), non-basic nitrogen (indole : pKa = -2.3) and ammonia (as exemplified by tert-butylamine : pKa = 10.8).

These experiments utilized Ethylflo/Naphthalene as the feed spiked to concentrations of 1, 2, 4, 8, 15, 25, and 50 ppmw nitrogen with the organo-nitrogen compound in the sequence quinoline, indole, tert-butylamine. At the conclusion of each experiment with a given organo-nitrogen species, i.e. following the test at the highest nitrogen concentration, the feed was changed to one that was nitrogen-free and the catalyst allowed to process this feed for a period of 10 - 14 days to examine the extent of recovery of saturation activity.

Introduction of nitrogen resulted in a decline in saturation activity. The yield of naphthalene, Fig. 12, increased with increasing concentrations of nitrogen, while the yields of tetralin and decalin decreased, Fig. 13 and Fig. 14, the latter nearly completely disappearing.

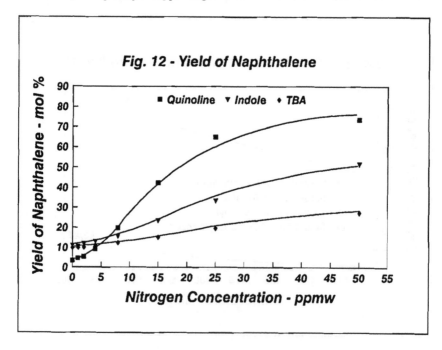

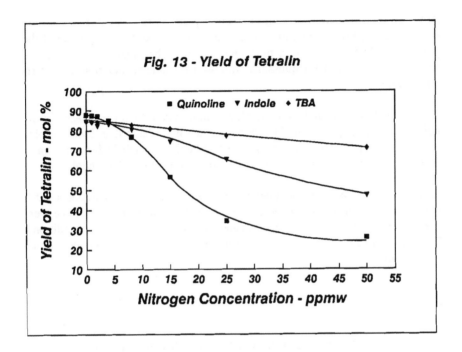

Fig. 13 - Yield of Tetralin

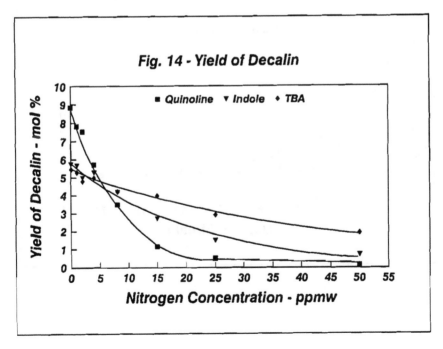

Fig. 14 - Yield of Decalin

Although the attenuation of saturation activity due to the presence of nitrogen compounds has been documented in the literature the extreme magnitude of the poisoning effect has not been fully appreciated. The greatest attenuation of activity was observed for quinoline, followed by indole, and finally by tert-butylamine.

Clearly, both basicity and structure play an important role in the potency of the poison. For example, quinoline which is much more basic than indole is also a much more potent poison, while tert-butylamine, which is far more basic than quinoline is not nearly as potent a poison, even less so than indole.

Denitrogenation was not examined during the course of the experiments but it would be reasonably safe to conjecture that neither quinoline nor indole were denitrogenated to any significant degree at these mild process conditions. Tert-butylamine may have been converted to a large extent to ammonia, which was the intent.

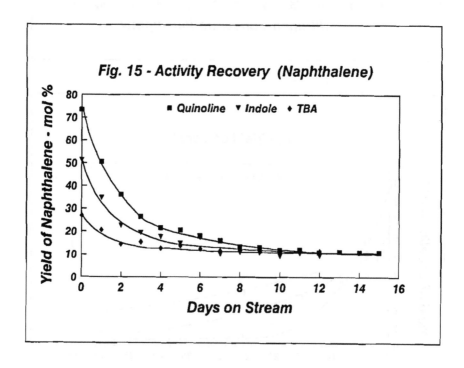

Fig. 15 - Activity Recovery (Naphthalene)

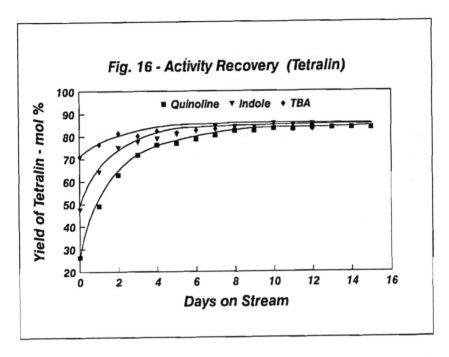

Fig. 16 - Activity Recovery (Tetralin)

Upon switching to a nitrogen-free feedstock saturation activity was recovered. The yield of naphthalene, Fig. 15, decreased while the yields of tetralin and decalin, Fig. 16 and Fig. 17, increased, with the yields of all three components approaching values observed prior to the introduction of the poisoning agents.

The poisoning was essentially completely reversible, as the catalyst regained saturation activity upon the removal of the poisoning agent from the feed, although a fairly long period of time, 10 - 14 days, was needed to effect the recovery. This implies that the poisoning action is due to an adsorption of the poison on the active sites of the catalyst and a blockage, or removal, of these sites from participation in the reaction. This conjecture is given further credibility by the fact that activity is slowly recovered following the removal of the poison as the adsorbed species desorbs and is flushed from the system.

Applying these concepts the poisoning action was modelled as an adsorption term appearing in the denominator of the overall rate constant consisting of an adsorption constant K_x, and a concentration of the poison, C_n, expressed as ppmw nitrogen, Fig. 18.

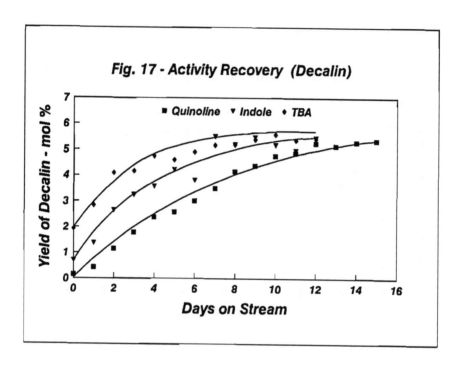

Fig. 17 - Activity Recovery (Decalin)

Fig. 18 - First Order Consecutive Reactions

Nitrogen Poisoning

$$k_1^* = k_1/(1+K_X C_N)$$

$$k_2^* = k_2/(1+K_X C_N)$$

$$k_1 = e^{(26.812-12825/T)}$$

$$k_2 = e^{(20.795-11492/T)}$$

Quinoline: $K_Q = 0.196$ Indole: $K_I = 0.0412$

Tert-butyl amine: $K_T = 0.0143$

The individual adsorption constants, K_Q, K_I, and K_T were determined by a nonlinear least squares fit to the data for the yield of naphthalene and were assumed to hold for the second reaction, the saturation of tetralin, as well as for the first. There is some justification for this assumption since it would be reasonable to assume that the same sites are involved for the saturation of either naphthalene or tetralin and that the adsorption of a poison on such a site would result in the same extent of attenuation of activity for each reaction.

The validity of such an assumption can be determined by whether the calculated yields of tetralin and decalin are reasonably close to those observed experimentally. If they are not then one must conclude that either the saturation of naphthalene and tetralin proceed on different sites or that the effect of the poison is different for the saturation of the two aromatics.

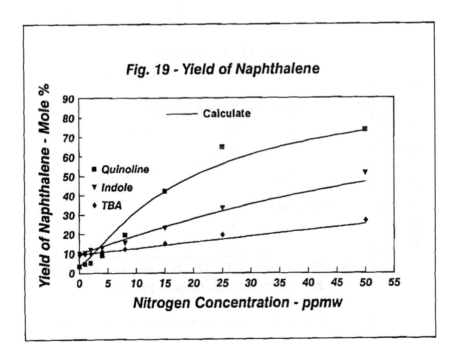

The above procedure resulted in an excellent fit to the data for all three components - naphthalene, tetralin, and decalin, Figs. 19 - 21. This lends considerable credibility to the assumption that saturation of naphthalene and tetralin proceed on the same sites which are susceptible to poisoning by the adsorption of organo-nitrogen compounds or ammonia.

The values of the adsorption constants reflect the strength of adsorption and the potency of the poison.   Quinoline, the most potent poison, has an adsorption constant of $K_Q = 0.196$, while indole, a much less potent poison, has an adsorption constant of $K_I = 0.0412$. Tert-butylamine, the least potent poisoning agent, exhibits the lowest value for the adsorption constant, $K_T = 0.0143$.

Both basicity and structure play an important role in the assessing the poisoning effect of an organo-nitrogen compound. Structure may play a role in the strength of the adsorption of the poison on the catalytic site quite apart from the role played by basicity, for example tert-butylamine is much more basic than indole yet it has a lower adsorption constant and exerts a much less severe poisoning effect.

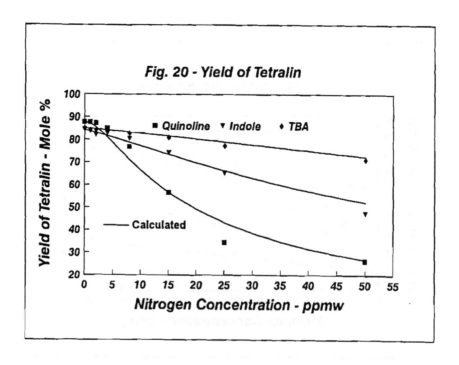

Fig. 20 - Yield of Tetralin

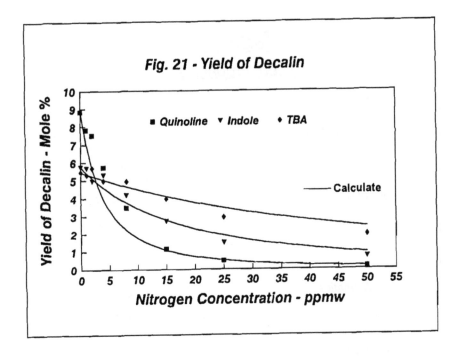

Fig. 21 - Yield of Decalin

Processing a Diesel Feedstock - Nitrogen Poisoning

To determine whether the phenomena observed for the processing of model aromatic compounds are also observed in the processing of a commercial diesel feed, an experiment utilizing a previously hydrotreated diesel feedstock was conducted. This feedstock had been hydrotreated to contain only 2.4 ppmw nitrogen and less than 5 ppmw sulfur and had an aromatics content of 23.4 vol % as determined by FIA. Additional properties are presented in Table I.

Table I

Feed Properties

API Gravity	=	37.7
Distillation, D-86		
IBP	=	362 deg F
50 vol %	=	468 deg F
End Point	=	631 deg F
FIA Aromatics	=	23.4 vol %
Sulfur	=	< 5 ppmw
Nitrogen	=	2.4 ppmw

This feed was processed over the same catalyst (although the process conditions were more severe with a pressure of 1000 psig and a temperature of 680 deg F) and the nitrogen concentration was increased via the addition of quinoline to levels of 10, 25, 50, 100, 200, and 400 ppmw as nitrogen.

The same attenuation of saturation activity was observed with the product aromatics concentration increasing with increasing concentration of quinoline in the feed, Fig. 22. Significantly, the greatest attenuation in activity occurred upon the initial introduction of quinoline (in the concentration range of 0 - 50 ppmw) followed by a much less pronounced deactivation as the concentration of the poison was increased further.

Additionally, it must be pointed out that the product did not contain nitrogen, even in the case when the feed contained 400 ppmw nitrogen as quinoline. In effect, what was observed was the poisoning effect of ammonia, the attenuation of saturation activity by organo-nitrogen compounds would be expected to be even greater.

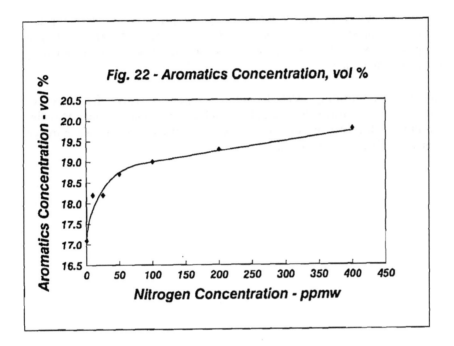

Conclusions

The saturation of di-aroimatics over Ni/Mo hydrotreating catalysts proceeds via a sequential mechanism with the saturation of the first aromatic ring followed by the saturation of the second.

The rate of saturation of a di-aromatic is dependent upon its structure, e.g. the saturation of naphthalene is 10 - 20 times faster than the rate of saturation of biphenyl. A difference of this magnitude in reactivities was not observed for the rates of saturation of the corresponding mono-aromatics, tetralin and cyclohexylbenzene.

Although a mono-aromatic is less reactive toward saturation than the parent di-aromatic the relative rates of saturation may vary widely, e.g. naphthalene undergoes saturation 20 - 40 times faster than tetralin while the rate of saturation of biphenyl is only 1 - 8 times greater than that of cyclohexylbenzene. The relative rates of saturation of di- and mono-aromatics are also dependent upon the structures of the two compounds under consideration. A simple "rule-of-thumb" that di-aromatics undergo saturation 10 times faster than mono-aromatics may be inaccurate and misleading.

Saturation activity is greatly inhibited by the presence of either organo-nitrogen compounds or ammonia. Both the type of organo-nitrogen compound, its structure, as well as its basicity are important in determining its potency as a poison.

Although hydrotreating catalysts are not especially active for aromatics saturation (as compared, for example, to noble metal catalysts) their activities are further attenuated by the presence of ammonia and/or organic nitrogen compounds in the process fluid.

References

1) McCulloch, D.C., and Edgar, M.D., "Higher Severity Diesel Hydrotreating", paper presented at the 1987 NPRA Annual Meeting, 29 - 31 Mar 87, San Antonio, Texas

2) Unzelman, G.H., "Diesel Fuel Quality - Refining Constrictions and the Environment", paper presented at the 1987 NPRA Annual Meeting, 29 - 31 Mar 87, San Antonio, Texas

3) Nash, R.M., "Meeting the Challenge of Low Aromatics Diesel", paper presented at the 1989 NPRA Annual Meeting, 19 - 21 Mar 89, San Francisco, California

4) Youngblood, D.J., "Industry Perspectives on Production of Low Sulfur/Low Aromatics Diesel Fuel", paper presented at the 1990 AIChE Annual Meeting, Nov 90, Chicago, Illinois

5) Davis, B.C., "Business Implications of New Fuel Regulations", paper presented at the 1990 AIChE Annual Meeting, Nov 90, Chicago, Illinois

6) McCarthy, K.M., and Felten, J.R., "Cost of Reducing Aromatics and Sulfur in Diesel Fuels", paper presented at the 1990 AIChE Annual Meeting, Nov 90, Chicago, Illinois

7) Zoller, J.R., Asim, M.Y., and Keyworth, D.A., "Hydrotreating for Production of Low Sulfur, Low Aromatics Diesel Fuel", paper presented at the 1990 AIChE Annual Meeting, Nov 90, Chicago, Illinois

8) Milam, S.N., Winquist, B.H.C., Murray, B.D., and Del Paggio, A.A., "Reduction of Aromatics in Diesel Fuel - A Low Pressure Two Stage Approach", paper presented at the 1991 AIChE Spring Meeting, Apr 91, Houston, Texas

9) Lee, S.L., and Jonker, R.J., "Aromatics Hydrogenation of Diesel Feedstocks", paper presented at the 1990 AIChE Annual Meeting, Nov 90, Chicago, Illinois

10) Shih, S.S., Mizrahi, S., Green, L.A., and Sarli, M.S., "Deep Desulfurization of Distillate Components", paper presented at the 1990 AIChE Annual Meeting, Nov 90, Chicago, Illinois

11) Kokayeff, P., "Aromatics Saturation Over Hydrotreating Catalysts - The Saturation of Naphthalene and Tetralin", paper presented at the 1992 AIChE Spring National Meeting, New Orleans, Louisiana, April, 1992

15 Production of Swedish Class I Diesel Using Dual-Stage Process

Barry H. Cooper, Peter Søgaard-Andersen, Peter Nielsen-Hannerup

Haldor Topsøe A/S
Nymøllevej 55, DK-2800 Lyngby
Denmark

INTRODUCTION

Regulations have been introduced in both Europe and the U.S. with a view to reducing hazardous emissions from diesel motors by limiting the content of sulfur and aromatic compounds in diesel.

In 1991, the Swedish Government proposed the introduction of three classes of diesel, each with a specific "energy tax". Environmental Class I diesel (SCI) specifies a maximum aromatic content of 5 vol% and a maximum sulfur content of 10 wppm. Class II diesel (SCII) limits aromatics to max. 20 vol% and sulfur to max. 50 wppm, whilst Class III (SCIII) has no limit on aromatics and a maximum of 0.05 wt% sulfur. The full specifications for the three classes are given in Table 1.

The difference in taxation levels for the three classes is considerable: Skr250/m^3 (corresponding to approx. $6/bbl) between SCII and SCIII, and a further Skr200/m^3 (corresponding to approx. $5/bbl) between SCI and SCII.

Topsøe has developed a two-stage process for low-aromatics diesel production at low partial pressures of hydrogen [1]. In the following, the application of this process concept is described for production of Swedish Class I diesel from North Sea gas oils.

279

Table 1: Specifications for Swedish Class I, II, III Diesel

	SCI	SCII	SCIII
Sulfur content max. wppm	10	50	500
Aromatic content max. vol%	5	20	-
Distillation IBP (min.) °C	180	180	-
10% (min.) °C	-	-	180
95% (max.) °C	285	295	340-370 *
Density @15°C kg/m³	800-820	800-820	820-860 *
Polyaromatic hydrocarbons, max. vol%	0.02	0.1	-
Cetane Index min.	50	47	43-46 *

* Depending on grade

PROCESS DESCRIPTION

The Topsøe two-stage process consists of the following main process sections:

*   Initial hydrotreating using a high activity base-metal catalyst.
*   Intermediate stripping to remove H_2S and NH_3.
*   Aromatics hydrogenation section using a sulfur-tolerant noble-metal catalyst.
*   Product stripping.

The process layout is shown in Figure 1. The purpose of the initial hydrotreating is to reduce the sulfur and nitrogen content in the diesel to a sufficiently low level to be able to achieve the required degree of aromatics removal in the aromatics hydrogenation section. The catalyst employed in this section would typically be a high activity NiMo catalyst such as TK 525. After intermediate stripping to remove H_2S and NH_3, the pretreated diesel is sent to an aromatics hydrogenation section containing a Topsøe noble-metal catalyst, TK 908. This type of catalyst makes it possible to achieve a high level of aromatics removal on diesel feedstocks containing up to several hundred ppm of sulfur at low operating pressures [2]. The sulfur content of the diesel is also reduced. A small amount of naphtha and light gases is produced in this section and is stripped together with H_2S in the product stripper.

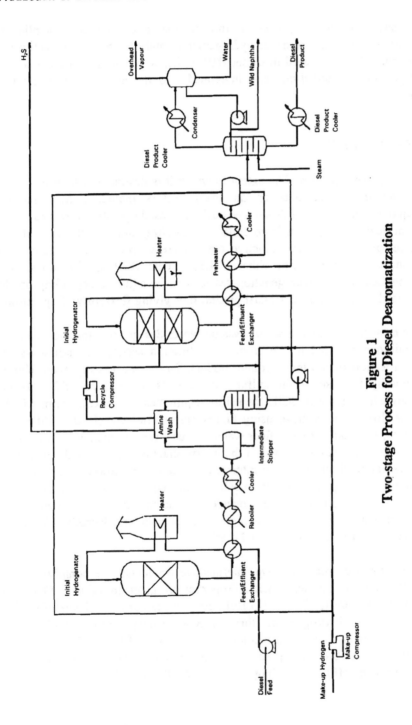

Figure 1
Two-stage Process for Diesel Dearomatization

An advantage of the process is the fact that it can be easily applied to existing units. The aromatics hydrogenation catalyst exhibits high activity at pressures typically used in diesel HDS units, which means that the second stage can be "added-on" and integrated into the existing desulfurization units.

TEST RESULTS

1. Feedstock Studies. The extent to which it is desirable to treat the diesel in the first stage depends on many factors including final product requirements, origin of diesel feedstock, proposed operating pressure and to what degree integration is possible with existing units. In the case of SCI diesel production, two attractive options are to process the intermediate product to comply with either SCIII or with SCII specifications. The throughput to the hydrogenation reactor can be adjusted depending on the demand for SCI and SCII/SCIII grades, thus giving flexibility of operation.

We have conducted a series of tests on North Sea gas oils to investigate the conditions required in the aromatics saturation section depending on the specification of the intermediate product. North Sea gas oils were chosen primarily because of the sulfur requirements for SCI and SCII. If the intermediate grade is SCII, the sulfur requirement of max. 50 ppm will be more difficult to achieve using higher sulfur crudes such as Middle East crudes, especially if the refiner wants to make use of existing hydrotreating units for the first stage. Even if the intermediate product is SCIII, it may be necessary to desulfurize the intermediate product to lower levels than the stipulated 500 ppm in order to achieve the required desulfurization in the aromatics section. A further consideration is the nitrogen content of the intermediate product, since a high nitrogen content adversely affects the aromatics saturation activity of TK 908.

The tests were made on two intermediate products: one complying with SCII grade and one with SCIII. Properties are given in Table 2. Operating conditions were LHSV = 1.5 l/l/h, reactor pressure = 50 kg/cm^{2}g, feed gas = 100% hydrogen once through. Reactor temperature was adjusted to give less than 5 vol% aromatics in the reactor effluent. The tests were performed in 50 cc bench-scale units (described in [2]). Aromatics analysis was made by HPLC according to the IP 391/90 method.

Table 2: Properties of Feeds Used in Second Stage Tests

Crude origin	Oseberg	Ekofisk
Diesel quality	SCIII	SCII
Density @15°C, kg/m³	832	820
Sulfur content, wppm	110	27
Nitrogen content, wppm	14	9
Aromatics content, IP 391/90 vol%		
Mono-	18.5	7.6
Di-	3.4	2.1
Tri-	0.3	0.0
Total	22.2	19.7
Aromatics content, ASTM D1319 vol%	24.6	20.0
Distillation, °C, ASTM D86		
IBP / 10%	192/214	192/210
50% / 90%	246/292	235/272
EP	315	292

The results of the two tests are shown in Fig. 2. It was found that it was necessary to operate at about 30°C higher temperature on the Oseberg feed to obtain less than 5% aromatics in the product.

2. Catalyst Stability. The results shown in Fig. 2 indicate that the catalyst system is quite stable. This was further demonstrated in an extended run on the Oseberg feed in which the catalyst was tested at a large number of conditions. The run was operated for a total of 5600 hours (230 days). Two of the test conditions were repeated during the run so that the catalyst stability could be measured. The results are shown in Fig. 3. In both tests, the unit was operated at LHSV = 1.5 l/l/h and total pressure = 50 kg/cm²g. At Condition 1, the temperature was originally (run hour 675-1075) adjusted to give 3.5% aromatics in the reactor effluent. When this condition was repeated 4000 hours later, the aromatic content of the product increased slightly to about 5%. The loss of activity represented by this increase in product aromatic content is very low (equivalent to 1°C/1000 hours). At Condition 2, the unit was operated with twice the H_2/oil ratio. A higher H_2/oil ratio results in a higher hydrogen partial pressure and a lower partial pressure of sulfur and nitrogen compounds. Both effects result in an

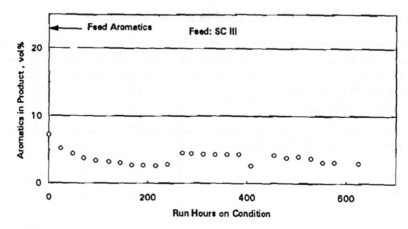

LHSV   = 1.5 l/l/h
Pressure   = 50 kg/cm²g
Temperature   = Base

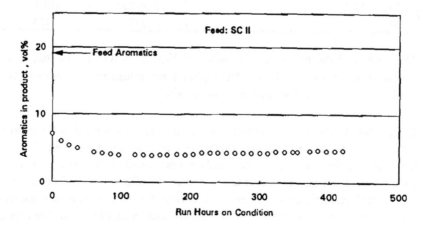

LHSV   = 1.5 l/l/h
Pressure   = 50 kg/cm²g
Temperature   = Base -30°C

Figure 2
Operating Conditions for SCI Production

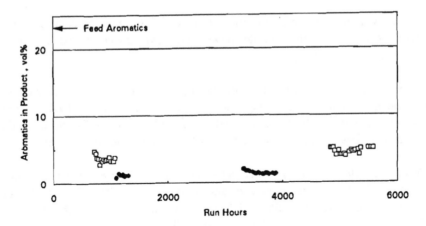

LHSV   = 1.5 l/l/h
Pressure   = 50 kg/cm^{2}g
□ : Condition 1, H$_2$/oil ratio = Base
● : Condition 2, H$_2$/oil ratio = 2 x Base

Figure 3
Catalyst Stability Test

increase in catalyst activity. The product aromatic content was 1.1% during the first period at Condition 2 (run hour 1075-1325) and 1.25% during the second period (run hour 3200-3900). This apparently represents an even lower rate of deactivation, but it is probable that the reaction is limited at Condition 2 by thermodynamic equilibrium.

3. Reaction Kinetics. The kinetics of aromatic saturation in diesel feedstocks are not straightforward. Polyaromatic compounds, i.e. compounds containing more than one aromatic ring, are saturated via consecutive reactions (e.g. 3-ring ⇒ 2-ring ⇒ 1-ring ⇒ naphthene) which are limited by thermodynamical equilibrium. There are few data on equilibrium constants available in the literature, but these suggest that the equilibrium for conversion of monoaromatics to naphthenes is the most favourable at the conditions normally seen in hydrotreaters. On the other hand, the monoaromatic species are less reactive than di- and triaromatics, so that the overall conversion is limited by the saturation of monoaromatics.

Several workers have made use of this fact and fitted experimental data for diesel feedstocks using simple power law expressions based on total aromatics content [3, 4].

In the dual stage process, most of the higher aromatics are converted to monoaromatics in the first stage (as seen from the resulting feeds to the second stage in Table 2). The kinetics of hydrogenation for these feeds might therefore be expected to be first order in total aromatics content.

Tests conducted at various space velocities show, however, that this is not the case. Product aromatic concentrations are given in Table 3 for tests run on the Ekofisk feed at LHSV = 1.5, 4, and 6 l/l/h. It is found that the data fit zero order kinetics better than first order (Fig. 4).

It is important to note that, at the conditions employed in these tests, there were no indications of equilibrium limiting the reaction. The good fit to zero order kinetics suggests that the reaction is inhibited by adsorption of unconverted aromatics, and that the kinetics would be better represented by a Langmuir-Hinshelwood - Hougen-Watson type rate equation.

In the conversion of a monoaromatic species to a naphthene, 3 moles of hydrogen are consumed per mole of aromatic compound. Tests made on the Oseberg feed at three different pressures show that the reaction rate is strongly dependent on hydrogen partial pressure (Fig. 5). Increasing the partial pressure of hydrogen from 35 to 60 results in a more than threefold increase in the activity of the catalyst. Based on these tests, it is estimated that the reaction order for hydrogen is between 2.5 and 3.

Table 3: Product Aromatic Content as a Function of LHSV

LHSV	1.5	4.0	6.0
Feed aromatics, vol%	19.7	19.7	19.7
Product aromatics, vol%	4.7	13.5	16.2

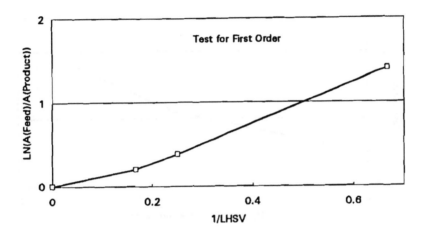

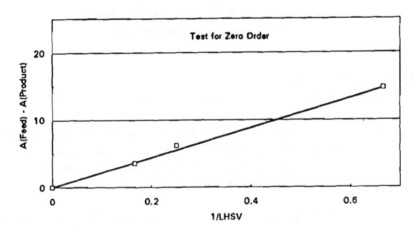

LHSV   = 1.5 1/l/h
Pressure   = 50 kg/cm²g
Temperature   = Base

Figure 4
Test for Order of Reaction

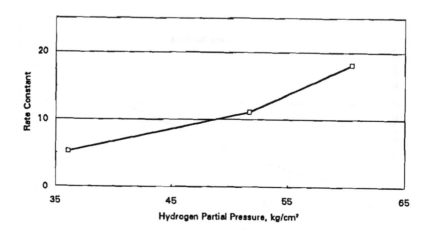

Figure 5
Influence of Hydrogen Partial Pressure on
Aromatics Saturation Activity

PROCESS CASE STUDIES

Process case studies have been carried out for two typical scenarios:

Case 1:   In an existing unit, a diesel is produced with a quality
           corresponding to the SCIII specification or poorer. SCI diesel
           is to be produced by adding a second stage with intermediate
           stripper and aromatics hydrogenation reactor.

Case 2:   A completely new two-stage unit is constructed to produce
           SCI diesel from straight-run gas oil, with the intermediate
           product meeting the SCII specification.

The first case is typical of many refineries producing diesel in Europe
today. By changing catalyst and/or operating conditions, many of these
units can produce SCIII or even SCII. The additional unit is envisaged as a
separate unit and the capacity need not match the capacity of the existing
unit. In this case the refiner will maintain the flexibility to process other
feedstocks in the existing unit and, through an intermediate storage tank, to
operate the two units independently.

The second case represents a situation where the refiner does not have any surplus desulfurization capacity and decides to install a dedicated facility to produce SCI with the inherent option to take out SCII as intermediate product if this should prove advantageous. If this option is exercised, it is necessary to include a separate, small fractionation section between the stages to adjust for the difference in 95 % cut point between SCI and SCII.

For the two cases, the following conditions have been assumed:

Production capacity: 20,000 bbl/day.
Feedstock, case 1:   SCIII Feed in Table 2.
Feedstock, case 2:   SCII Feed in Table 2.
Hydrogen: No hydrogen plant foreseen;
                       cost of hydrogen = $1.41US\$/10^3SCF$.

Catalyst, case 1:   No change of catalyst in existing hydrotreater;
                       TK 908 in second stage.
Catalyst, case 2:   TK 525 in first stage; TK 908 in second stage.

The production costs have been calculated for the two cases and the key figures are given in Table 4. Other operating costs cover cost of fuel, utilities, manpower, catalyst, etc.. Capital costs are taken as 25 % of investment cost.

The calculated production costs demonstrate that production of SCI by the Topsøe two-stage process is feasible in both cases.

Table 4. Production Costs (1000 US$/year)

Case	1	2
Hydrogen	2,500	4,100
Other operating costs	6,000	8,700
Total operating costs	8,500	12,800
Capital costs	4,800	10,000
Total costs	13,300	22,800
Costs, US$/bbl	2.02	3.45

Taking into consideration the difference in taxation levels, sufficient margin is left to cover additional costs to dedicated storage and distribution facilities. The two cases should not be seen as alternatives. There can be many reasons for choosing one or the other solution: logistics, market requirements, crude supply, outlet for the diesel fraction between SCIII and SCII/SCI, layout of existing equipment, etc. For both cases, however, there is an economic incentive to produce SCI rather than the intermediate products, SCII or SCIII.

REFERENCES

[1]   "Topsøe's Process for Improving Diesel Quality", P. Søgaard-Andersen, B.H. Cooper, P.N. Hannerup, NPRA AM, New Orleans March 1992, Paper AM-92-50.

[2]   "Diesel Aromatics Saturation - A Comparative Study of Four Catalyst Systems", B.H. Cooper, A. Stanislaus, P.N. Hannerup, ACS Preprints, Div. of Fuel Chemistry, Vol. 37, No. 1, (1992), p. 41-49.

[3]   "Upgrading of Middle Distillate Fractions of a Syncrude from Athabasca Oil Sands", M.F. Wilson, J.F. Kriz, Fuel, 63, (1984), p. 190-196.

[4]   "Kinetics of Aromatics Hydrogenation of Bitumen-Derived Gas Oils", S.M. Yui, E.C. Sanford, Can. J. Chem. Eng., 69,(1991), p. 1087-1094.

16 Shell Middle Distillate Hydrogenation Process

J.P. Lucien[+], J.P. van den Berg[*], G. Germaine[#],
H.M.J.H. van Hooijdonk, M. Gjers[@] and G.L.B. Thielemans

Shell Internationale Petroleum Mij.
The Hague, The Netherlands

[+] *Companie Rhenane de Raffinage Reichstett, Reichstett Vendenheim, France*

[#] *Shell Recherche SA, CRGC, Grand-Couronne, France*

[@] *Shell Raffinaderi AB, Gothenburg, Sweden*

ABSTRACT

The strive towards cleaner environment has lead to low sulphur specifications for middle distillate fuels. In addition compositional specifications are presently debated. Thus, to meet future emissions standards regarding, specifically, particulates emissions, the motor industry calls for improved automotive gasoil quality. Although automotive gasoil quality affects emissions from diesel engines it is considered less influential than engine design and maintenance. Sulphur, density and cetane number are the fuel properties having the greatest influence on diesel engine emissions although also aromatics and endpoint specifications have been defined in environmentally adopted government initiatives.

---

[*] to whom correspondence should be addressed

This paper reviews the options which are available to tackle these new requirements. The high severity single stage concept (using conventional mixed sulphides catalysts) will be discussed in its potential to meet more severe product requirements as well as in terms of its limitations, especially at the point of aromatics saturation and cetane upgrading. Furthermore it is shown that the option of severe hydrotreating followed by hydrogenation with conventional noble metal catalysts is preferred if deep aromatics saturation is aimed at. However, this conventional two stage concept has limitations with respect to heaviness and sulphur and nitrogen content of feedstocks.

The new Shell Middle Distillate Hydrogenation (SMDH) technology, applying a (semi) two stage approach based on the Shell developed hydrogenation catalyst is presented. The SMDH process will be discussed in its potential to break the limitations of the conventional options. The new catalyst is crucial in this process and allows a highly integrated mode of operation. It is characterised by a high tolerance to sulphur and nitrogen, a high stability and robustness against sulphur and nitrogen upsets.

A number of applications of this novel process will be discussed, especially the production of environmentally adapted AGO from SR-LGO in Sweden and the upgrading of cycle oils in premium quality middle distillates.

The first commercial application of the SMDH technology at the Shell Refinery in Gothenburg will be reported. The unit produces on spec Swedish Class 1 diesel (< 5 %v aromatics) with an excellent yield.

INTRODUCTION

Trends in product quality specifications

Emissions regulations and, related to that, fuel quality requirements have shown a rapid development towards lower emissions and cleaner fuels over the past few years [1].

Until recently most legislation with respect to diesel fuel quality was focused on sulphur content. Within the European Community, for example, the Environmental Council adopted rules imposing a maximum S limit of 0.2 %m

Table 1 *Gasoil product quality specifications*

A. Swedish Diesel Fuel Classifications. Urban Diesel Grades I/II

	Current *	Class II	Class I
S (%w)	0.1	0.005	0.001
Aromatics (%v)		20	5
Poly-aromatics (%v)		0.1	0.02
Distillation:			
IBP (degC)		180	180
10 %v ASTM (degC)	180 **		
95 %v ASTM (degC)		295	285
Density (g/l)	800-840 **	800-820	800-820
Cetane index (D4737)	46	47	50
Tax ($/m3) ***	209	173	144
Tax incentive ($/m3)		36	65

* Current practice to avoid S tax
** Winter Grade
*** ROE 1 US$ = 7 SKr

B. CARB Reference Fuel Specifications

	< 50000	> 50000
Ref.Capacity (bpd)		
S (ppmw)	500	500
N (ppmw)	90	10
Aromatics (%v)	20	10
Polyaromatics (%v)	4	1.4
Cetane Number clear	47	48
Density (kg/m3)	830-860	830-860
FBP ASTM (degC)	349	349
Visc (40 degC, cSt)	2.0-4.1	2.0-4.1

for all GO and diesel fuels by 1 October 1994 and of 0.05 %m for diesel fuel by 1 October 1996.

In addition to sulphur reduction a number of countries are introducing compositional specifications in the belief that this will enhance emissions performance. In parallel a continuous debate still is in progress to establish the most critical parameter(s) [2]. In legislation, nevertheless, examples of setting minimum cetane number/index levels, maximum density values, maximum aromatics contents and reduced final boiling points (FBP). As an example the Swedish and Californian (CARB) specifications are listed in Table 1.

The dual challenge

In parallel to the call for improved quality is the need to satisfy a growing demand for AGO, which will necessitate the full utilisation of conversion streams as well as straight run gasoil components. Important to note that cracked GO blending components, ex thermal (TCGO) and catalytic cracking (LCO/HCO), may represent up to 30 % of the total GO pool in a complex refinery. Simultaneously these compounds are found to be highest in sulphur and density and lowest in cetane quality. In addition they are also highest in nitrogen and aromatics contents (Figure 1).

Thus refineries nowadays are confronted with a dual challenge which consists in the requirement of manufacturing premium AGO, fulfilling higher quality specifications, in increasing quantities requiring the upgrading of more refractive feedstocks. At present refineries are already investing heavily to achieve the AGO sulphur levels of 0.05 %m (within EEC by October 1996). In parallel new process schemes and/or technologies are being developed to enable them to meet future specifications on e.g. cetane quality, density, boiling range and colour stability.

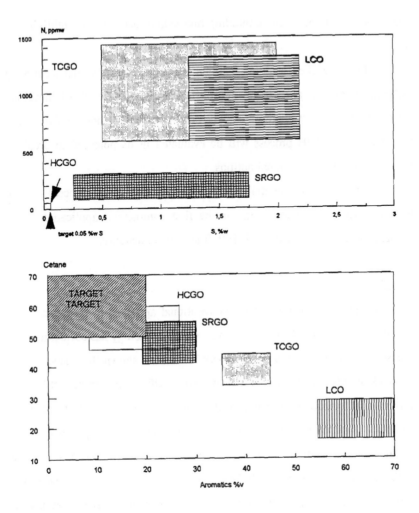

Figure 1: *Feedstock characteristics for refinery gasoil pool*

In this paper a number of options to tackle the increasing fuel quality requirements will be discussed. The single stage concept (using conventional mixed sulphides catalysts) will be analysed in its potential to meet more severe product quality requirements. Its limitations, however, with respect to e.g.

aromatics saturation, cetane upgrading and colour stability are highlighted as well.

The new Shell Middle Distillate Hydrogenation (SMDH) technology, applying a (semi) two stage approach, based on the Shell developed hydrogenation catalyst is presented. The crucial role of the new catalyst will be indicated. The SMDH process will be evaluated in its potential to break the limitations of the conventional options.

A number of potential applications of this novel process will be discussed with special attention for the first commercial application of the SMDH process at the Swedish Shell Refinery in Gothenburg.

PROCESS OPTIONS

In recent years the potential of the conventional approach to manufacture low sulphur/low aromatics AGO using single stage technology has been extensively discussed [3-9]. In many of these papers it has been shown that, dependent on feedstock quality, in general the low sulphur specifications can be met by using conventional hydrotreating catalysts at higher severity [3-6,8]. With respect to aromatics saturation, hydrodenitrogenation and related product quality specifications (e.g. colour stability) we are faced with limitations of single stage solutions [3,5-9]. This especially holds for cases where cracked feedstocks are included in the gasoil pool and/or where stringent aromatics specifications have to be met [5-8].

Mixed metal-sulphide catalysts in single stage units

Co-Mo mixed suphides in general show excellent hydrodesulphurisation [10-11] properties. For applications where an increased hydrogenation activity is required, like hydrodenitrogenation, Ni-Mo and/or Ni-W type catalyst will be the preferred systems [9]. As a typical example in Figure 2, HDS and HDN

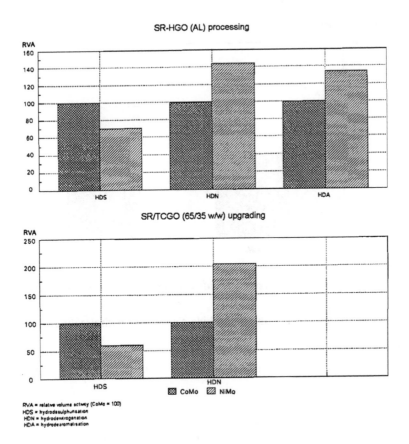

Figure 2: *Effect of catalyst composition on hydrotreating activities HDS, HDN and HDA activity compared*

activities for CoMo and NiMo catalysts are compared for a straight run heavy gas oil (SR-HGO) and a thermally cracked GO containing blend (SR/TCGO), respectively.

Also on aromatics saturation the Ni-containing catalyst formulations are the preferred catalysts for single stage deep-hydrogenation [6,8,10]. As a typical example in Figure 3 the aromatics saturation performance of a CoMo

aromatics conversion %

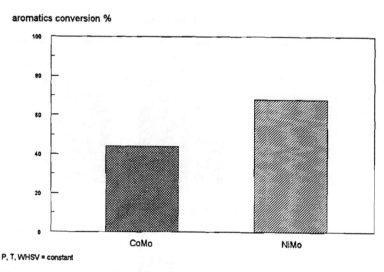

P, T, WHSV = constant

Figure 3: *Selection of mixed sulphide catalysts for aromatics saturation*

type catalyst is compared with Ni-sulphide catalysts in hydrotreating of a LCO
containing feedstock.

Feedstock refractiveness (S,N content, boiling range etc.) determines

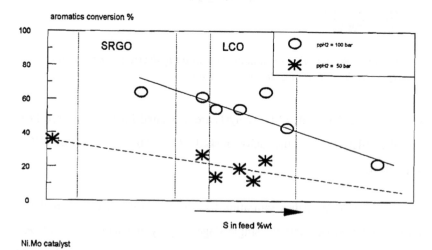

Ni.Mo catalyst

Figure 4: *Effect of hydrogen partial pressure and feed refractiveness on*
shown

the maximum achievable degree of aromatics saturation to a large extent. As is shown in Figure 4 for a typical LCO feedstock hydrogen partial pressures up to 100 bar and large reactor volumes are required to reach an average level of 50% aromatics saturation.

Furthermore aromatics saturation is a reversible reaction favoured by low temperatures [10]. Consequently even at high hydrogen partial pressures the temperature operating window is limited. For example Figure 5 shows the shift in operating window with the hydrogen partial pressure as typically applied for different feedstock refractiveness.

Due to the thermodynamic limitations for single stage saturation of aromatic structures improvement of other product quality parameters is limited as well. Limitations in cetane upgrading and colour stability after deep hydrotreating have been reported recently [5-8]. As an example in Figure 6 the limitation on product colour performance is shown in relation to the limitations

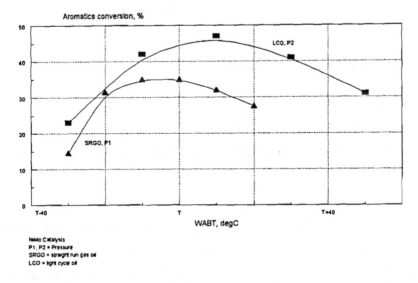

Figure 5: Limitations of the single stage concept. Temperature operating window for aromatics saturation

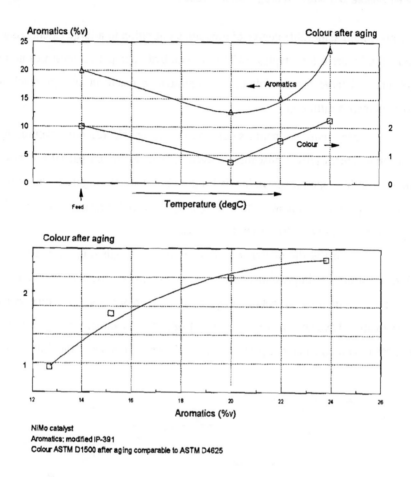

Figure 6: *Limitations of the single stage concept -- Temperature operating window for product colour specifications*

in aromatics saturation in deep-HDS of a SR-LGO using a commercial NiMo-type catalyst.

It is concluded that for single stage HDS units the maximum reactor outlet temperature (ROT) is constrained when stringent product quality specifications have to be met. Adding to this the more rapid catalyst deactivation under deep HDS conditions, this observation leads to seriously constrained catalysts cycle lengths. Considering that, on the contrary, longer operation cycles nowadays are desired in refineries, it becomes clear that deep HDS only is possible at the expense of large reactor volumes and/or increased hydrogen partial pressures. Although feasible, such options may require high capital investments while offering only limited flexibility, particularly when cracked feedstock components are included in the HDS feed diet.

Conventional two stage configurations for aromatics saturation

As demonstrated above the operating conditions required for deep aromatics saturation, using mixed sulphide catalysts, rapidly run into thermodynamic limitations. Especially with the more refractive feedstocks, ie. containing cracked feedstock components, a limited degree of aromatics saturation is the consequence.

An alternative route to break these limitations is found in the application of dedicated hydrogenation catalysts. Noble metal catalysts allow deep hydrogenation at considerably lower temperatures thus avoiding the thermodynamic limitations encountered with the mixed sulphide catalysts.
In a number of examples, as developed over the years [12-16], application of noble metal catalysts indeed showed the potential of this route.

Clear disadvantage of this route, however, is the sulphur and nitrogen sensitivity of the conventional noble metal catalyst. Consequently a two stage operation is required with a deep hydrotreating in the first stage followed by extensive intermediate product separation and stripping to assure a S and N reduction in the first stage product typically below 5 and 2 ppmw, respectively [17].

Second stage hydrogenation then allows reduction of aromatics levels down to values ranging from typically 10-15 %v (smoke point improvement) to ppmw level (solvent production). Based on this technology, e.g. within the

Shell group, several hydrogenation units are in commercial operation on the production of low aromatics solvents, medicinal oils, jet fuels with improved smoke points and sometimes on cetane improvement of SR-LGO. Only by applying careful feedstock handling, appropriate hydrogen partial pressures and high activity catalysts, acceptable cycle lengths can be achieved.

As a typical example for conventional hydrogenation technology in Table 2 some data on kero hydrogenation for smoke point improvement are listed. Although mild conditions are sufficient to obtain deep hydrogenation (> 80% saturation) very low feedstock S and N contents are essential to assure a stable catalyst performance and low temperature requirements. With increasing feedstock S and N contents a rapid increase in temperature requirement is found which may prevent deep saturation due to thermodynamic limitations.

It is because of this sensitivity that application of conventional hydrogenation technology is essentially limited to the hydrogenation of light reactive feedstocks. Application for saturation of more refractive feedstocks,

Table 2 *Conventional aromatics saturation, Low aromatics kerosene saturation*

	Feedstock	Product
Yield (%w)		> 99.5
CCH2 (%w)		0.74
Density (kg/m3)	819	809
ASTM-D86 (degC)		
10 %v	204	204
50 %v	229	225
90 %v	268	265
S (ppmw)	2	< 1
N (ppmw)	1	< 1
Tot Aromatics (%v)*)	26.0	5.0
Smoke Point	20	29

*) modified IP-391
CCH2 = chemical hydrogen consumption
Smoke Point - ASTM D1322

e.g. LCO, would require severe first stage hydrotreating conditions (PPH2 in the range of 70-100 bar) to meet the second stage feedstock sulphur and nitrogen specifications.

Integrated two stage approach with S,N tolerant hydrogenation catalysts

With the introduction of a new family of sulphur and nitrogen tolerant zeolite based noble metal catalysts Shell [18] developed a higly integrated two stage middle distillate hydrogenation technology. This new "Shell Middle Distillates Hydrogenation" (SMDH) process allows deep saturation of GO range feedstocks at relatively low hydrogen partial pressures.

High feedstock S and N contents, up to 1000 and 50 ppmw respectively, are tolerated by the second stage hydrogenation catalyst while maintaining its activity and stability. In addition this type of catalyst shows an extremely attractive feature in its potential as hydrodecyclisation catalyst [19] resulting in a significantly higher cetane upgrading as compared to conventional hydrogenation catalysts. Consequently this new process is very well suited for saturation and upgrading of refractive feedstocks including cracked gas oils (see below).

Due to the high catalyst activity and stability only relatively mild conditions are required for deep desulphurisation and deep aromatics saturation [20]. In addition the catalyst robustness against sulphur and nitrogen contaminants allows a high degree of process integration thus decreasing capex costs to a large extent. The first commercial application of this technology was started up at the Shell Refinery at Gothenburg (Sweden) in August 1992.

More recently other (integrated) two stage process line-ups for manufacturing low aromatics diesel fuels have been proposed [21-26] but no commercial applications have emerged yet.

THE SHELL MIDDLE DISTILLATE HYDROGENATION PROCESS

In the development of the new zeolite supported noble metal catalyst (S-704) this catalyst was found to be extremely robust against high sulphur and nitrogen in the feed, this even at moderate pressures and high space velocities. In addition it is found that the catalyst can be easily rejuvenated in-situ upon severe S and/or N contamination. It is on the basis of this feature that a drastically simplified two stage process concept could be developed avoiding the conventionally required catalyst protections (deep stripping, guard-beds, etc.) against S and/or N contamination. This resulted in the highly integrated Two Stage process line-up, as shown in Figure 7.

Catalyst features

As stated above the main feature of the new S-704 noble metal catalyst is its robustness against sulphur and nitrogen contamination. This not only results in a high tolerance to feedstock S and N contents but also allows the

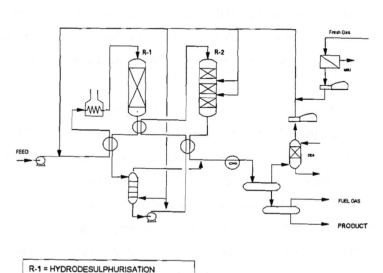

R-1 = HYDRODESULPHURISATION
R-2 = HYDROGENATION

Figure 7: *Process scheme of the Shell Middle Distillate Hydrogenation Process -- The Gothenburg line-up*

catalyst regeneration after S and/or N upsets. In addition the S-704 catalytic performance attributed to its specific formulation allows the following benefits:

- The catalyst shows an excellent stability. In Figure 8 the catalyst performance on a HT-LCO feedstock is shown. After an initial stabilisation period the catalyst deactivation rate is shown to stabilise at values significantly below 1 °C/1000 hrs.
- The S-704 catalyst is found to be highly selective for aromatics saturation. In Figure 9 it is shown that indeed maximum aromatics saturation can be reached while still avoiding any significant cracking. On the other hand, as far as cracking occurs the only by-product consists of heavy naphtha with low sulphur and nitrogen contents which can be routed directly to the platformer

The S-704 shows an attractive ring opening activity, resulting in a considerable additional cetane improvement on top of the cetane

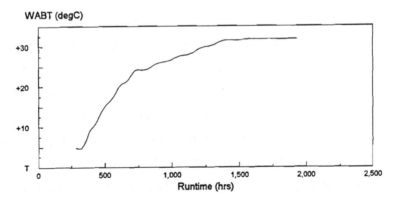

Feed: HT-LCO ex NiMo
Temp. requirement for 2 %v aromatics

Figure 8: *S-704 performance on HT-LCO hydrogenation Catalyst stability*

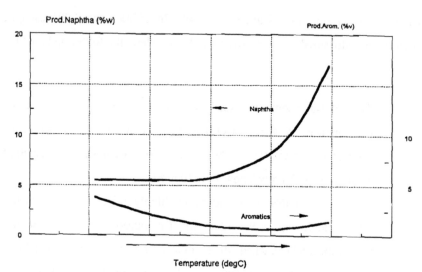

Figure 9: *Product selectivity in LCO upgrading using the two stage*
SMDH process -- Aromatics saturation down to Swedish Class 1 level

improvement due to aromatics saturation only [19]. This feature is especially attractive in case of feedstocks with high concentrations of ring components.

The potential of the S-704 catalyst for cetane upgrading is clearly shown in Figure 10. In this figure the relative upgrading over the two stages of the SMDH process for three product quality parameters is plotted. A typical LCO is used as feedstock. Over the first stage, containing a commercial mixed metals sulphide catalyst, the deep desulphurisation is shown while only a partial aromatics saturation and, related to that, cetane upgrading is found. Over the second stage, however, containing the new S-704 catalyst, aromatics saturation is virtually complete and a dramatic further cetane upgrading is observed. In Table 3 the data show that using the two stage SMDH technology, indeed, starting with refractive feedstocks like pure cracked feed components like LCO highly valuable middle distillates can be produced fulfilling most stringent product quality constraints, as specified in various legislations (e.g. Table 1).

Process characteristics

The unit essentially consists of two reactors operated at virtually the same pressure (Figure 7). In the first stage feed pretreating under rather moderate conditions results in sulphur and nitrogen reduction to levels acceptable for the second stage catalyst. The first stage gas/liquid effluent is separated in the intermediate high pressure stripper and the liquid phase fed into the second stage hydrogenation reactor. By tuning the second stage operating temperature the level of aromatics saturation can be easily controlled. The high pressure off-gas is purified by amine treating before recycling to the first and second reactor.

Based on the high S,N tolerance of the S-704 catalyst a maximum process integration is possible allowing the following benefits:

* Both reactors are run under mild operating conditions (P,T).

* Single separator system.

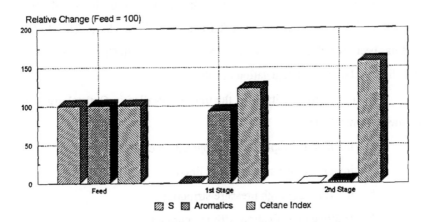

Figure 10: *The integrated two stage SMDH process - LCO upgrading*

* Maximum heat integration

* Maximum process flexibility since the product saturation level can be simply controlled by adjusting the second stage reactor temperature.

* The duty of the first and second reactor can be easily adjusted by balancing the first and the second stage reactor temperatures, e.g. a lower first stage temperature results in a higher S,N load for the second stage catalyst and thus results in a higher second stage temperature requirement.

Overall, due to the catalyst robustness, the high degree of process integration and the mild operating conditions required (low pressure and temperature), the SMDH process is found to be highly cost effective in the upgrading of low quality gasoils to automotive gasoils, fulfilling most stringent product specifications.

First commercial SMDH application at the Shell Raffinaderi at Gothenburg

In August 1992 the first commercial application of the SMDH process was successfully started at the Shell Raffinaderi at Gothenburg. This 1500 t/sd unit has been installed to supply the Swedish market with Class 1 type AGO, as specified by the Swedish legislation (Table 1).

Being the 'first-of-its-kind' the unit as installed in Gothenburg has the following new features:

* the unique proprietary hydrogenation catalyst
* integrated two stage process line-up
* intermediate high pressure stripping
* hydrogen purification with membranes

Table 3 *LCO upgrading via two stage deep hydrogenation*

	FEED	PRODUCT		
Yield				
Fuel gas, %w	-	0.1		
C3 + C4, %w	-	3.0		
Naphtha (180-), %w	-	12.8		
180+, %w	100	28.9 (K) + 55.2 (GO)		
CCH2 (%w) *)		3.75		
		Naphtha	Kero	Gasoil
Properties				
Specific Gravity	0.912	0.757	0.803	0.845
Sulphur, ppmw	21100	<1	<1	3
Nitrogen, ppmw	710	<1	<1	<1
ASTM 50%v, degC	249	125	186	254
Total aromatics, %v **)	65	1.5	1.6	1.4
Tri+ aromatics, %v	9	-	<0.05	<0.05
Smoke point, mm ***)	-	-	32	-
Cetane Index ****)	30.0	-	38	49
Cetane Number	-	-	-	51

*) CCH2 = chemical hydrogen consumption
**) modified IP-391
***) Smoke point - ASTM D1322
****) ASTM D4737

Table 4 *Class 1 and 2 AGO production for the Swedish market. First commercial application*

	Feedstock	Class 2	Class 1	Specs
CCH2 (%w) *)		0.55	0.95	(Class 1)
Yield (%w) - GO		97	93	
- Naphtha		3	7	
Density (kg/m3)	820	806	801	800-820
S (ppmw)	4500	15	5	< 10
N (ppmw)	10	< 1	< 1	
Tot Arom (%v) **)	30	18	4.5	< 5
ASTM D86 (degC)				
10 %v	200	198	195	
50 %v	235	233	231	
95 %v	285	285	285	< 285
Colour ***)			L 0.5	
Cetane Index ****)	50	55	57	> 50

*) chemical hydrogen consumption
**) modified IP-391
***) Colour ASTM D1500 after ageing comparable to ASTM D4625
****) Cetane Index - ASTM D4737

A full process scheme is shown in Figure 7. In addition to the features already discussed above the membrane unit for hydrogen purification allows a maximisation of the hydrogen partial pressure while minimising operating pressures.

The performance of the SMDH process when running on Statfjord SR-LGO for the production of Swedish type AGO (Table 1) is shown in Table 4. It is clear that all specifications are easily met while only mild hydrotreating severities are required. On top of that it should be noted that, even at deep saturation, the gas make is extremely limited.

The unit is found to be highly flexible towards producing Class 1 or Class 2 AGO when required. A simple adjustment of the second stage reactor temperature is sufficient, assuming sufficient hydrogen availability, to switch from Class 2 to Class 1 production.

Comparing the properties if the Statfjord SR-LGO feedstock, used in the Gothenburg unit (Table 4), with those of more refractive feedstocks, e.g. a typical LCO feedstock as listed in Table 3, it is clear that the duty for the S-704 catalyst in case of the Gothenburg unit is a relatively easy one. Consequently for the Gothenburg unit a long catalyst cycle time for the noble metal catalyst is envisaged.

CONCLUSIONS

The rapidly developing environmental legislation becomes more and more challenging for present and future oil refining and AGO manufacturing. It is shown that conventional single stage technology, using mixed metal sulphide catalysts, can be applied to meet more stringent product S (and N) specifications. Other specifications like cetane quality, aromatics content etc., however, show that the single stage concept is limited in its maximum severity due to thermodynamic limitations in aromatic saturation.

Recent developments in hydrogenation technology for middle distillate upgrading has resulted in the introduction of a new Shell proprietary S,N tolerant hydrogenation catalyst. The following features of this new zeolite based noble metal catalyst have been listed:

* High activity and stability on aromatics saturation at low pressure
* High sulphur and nitrogen tolerance
* Easy in-situ regenerability after S,N upset
* Additional ring opening activity (cetane upgrading)
* Low cracking activity

It is on the basis of this new type of catalysts that a new two stage process for deep hydrodesulphurisation/deep hydrogenation has been developed.

The first commercial application of this two stage deep AGO hydrogenation technology, i.e. the Shell Middle Distillate Hydrogenation Process, at Shell Raffinaderi at Gothenburg, is successfully started up in August 1992. In a listing of the product qualities obtained it is shown that the Class 1 and 2 AGO specifications, as listed in the Swedish legislation, are easily met. Given the relatively light duty for the S-704 when processing Statfjord SR-LGO a long catalyst life cycle is envisaged.

Due to the catalyst robustness, the high degree of process integration and the mild operating conditions required (low pressure and temperature), the SMDH process is found to be an excellent and competitive technology for the upgrading of low quality gasoils to automotive gasoils, fulfilling most stringent product specifications.

REFERENCES

1. Concawe, 1992. Motor vehicle emission regulations and fuel specifications - 1992 update. Report No 2/92

2.   M. Booth, J.M. Marriott and K.J. Rivers, 1993. Diesel Fuel Quality in an Environmentally Conscious World. Second Inst.Mech.Eng.Int.Seminar on Fuels for Automotive and Industrial Diesel Engines", April 6-7, 1993, London.

3.   R.M. Nash, 1989. Process conditions and catalyst for low-aromatics diesel studied. Oil gas J., May 29: 47-56

4.   A.J. Suchanek, 1990. Catalytic routes to low-aromatics diesel look promising. Oil Gas J., May 7: 109-119

5.   J.A. Anabtawi and S.A. Ali, 1991. Effects of catalytic hydrotreating on light cycle oil fuel quality. Ind. Eng. Chem. Res., $\underline{30}$: 2586-2592

6.   S.L. Lee and R.J. Jonker, 1991. Aromatics hydrogenation of diesel feedstocks. AKZO catalyst symposium 1991 - Hydroprocessing, June 1991, Scheveningen, 25-46

7.   T. Tatsuka, Y. Wada, H. Suzuki and S. Komatsu, 1991. Colour degradation of diesel fuel in deep desulphurisation. 1991 NPRA Annual Meeting, March 17-19, 1991, San Antonio, Paper AM-91-39

8.   S.L. Lee and M. de Wind, 1992. Single and dual stage hydrotreating can improve diesel cetane. Oil Gas J., Aug 17: 88-90

9.   T.C. Ho, 1988. Hydrodenitrogenation catalysis. Catal. Rev.-Sci. Eng., $\underline{30}$ (1): 117-160.

10.   M.J. Girgis and B.C. Gates, 1991. Reactivities, reaction networks and kinetics in high pressure catalytic hydroprocessing. Ind. Eng. Chem. Res., $\underline{30}$: 2021-2058

11.   R. Prins, V.H.J. de Beer and G.A. Somorjai, 1989. Structure and function of the catalyst and the promotor in Co-Mo hydrodesulfurisation catalysts. Catal. rev.-Sci. Eng., $\underline{31}$ (1-2): 1-41

12.   1969. Hydrogenation of aromatic hydrocarbons. US 3.592.758 to UNOCAL

13.   S.M. Kovach and G.D. Wilson, 1974. Selective hydrogenation of aromatics and olefins in hydrocarbon mixtures. US 3.943.053 to Ashland Oil

14.   R.C. Schucker and K.S. Wheelock, 1984. Process for the hydrogenation of aromatics. US 4.469.590 to Exxon Research

15.   Shell SPI technology, 1986. Hydrocarbon Processes, September : 87

16.   J.P. Peries, A. Billon, A. Hennico and S. Kressmann, 1991. IFP deep desulphurisation and aromatics hydrogenation on straight run and pyrolysis middle distillates. 1991 NPRA Annual Meeting, March 17-19, 1991, San Antonio, Paper AM-91-38

17. M.Y. Asim, D.A. Keyworth, J.R. Zoller, F.L. Plantenga and S.L. Lee, 1990. Hydrotreating for ultralow aromatics in distillates. 1990 NPRA Annual Meeting. March 25-27, 1990, San Antonio, Paper AM-90-19

18. J.K. Minderhoud and J.P. Lucien, 1988. Process for the hydrogenation of hydrocarbon oils. EP 303.332 to SIRM

19. J.P. Lucien and G.L.B. Thielemans, 1991. Hydrodecyclization Process. EP 512.652 to SIRM

20. S.N. Milam, B.H.C. Winquist, B.D. Murray and A.A. DelPaggio, 1991. Reduction of aromatics in diesel fuel. A low pressure two stage approach. Spring National Meeting of AIChE, April, 1991.

21. E.C. Haun, G.J. Thompson, J.K. Gorawara and D.K. Sullivan, 1990. Two-stage hydrodesulphurisation and hydrogenation process for distillate hydrocarbons. US 5.114.562 to UOP

22. J.W. Reilly and G.L. Hamilton, 1992. Production of diesel fuel by hydrogenation of a diesel feed. WO 92/16601 to ABB Lummus-Crest.

23. G.L. Hamilton and A.J. Suchanek, 1990. Flexible new process converts aromatics in variety of diesel feedstocks. Oil Gas J., July 1: 55-56

24. G.L Hamilton, E.L Graniss, L.J. Scotti and A.J. Suchanek, 1992. SYNSAT produces high quality diesel at low pressure. JPI Petroleum Refining Conference 1992, Nov 1992, Tokyo, Paper 2.2

25. P. Sogaard-Andersen, B.H. Cooper and P.N. Hannerup, 1992. Topsoe's process for improving diesel quality. 1992 NPRA Annual Meeting, March 22-24, 1992, New Orleans, Paper AM-92-50

26. D. Eastwood, C. Tong and S.E. Yen, 1992. Integrated hydrotreater/aromatic saturator process for the production of oils with reduced aromatics and polynuclear aromatics. 1992 NPRA Annual Meeting, March 22-24, 1992, New Orleans, Paper AM-92-63

17 A Comparative Study of Catalysts for the Deep Aromatic Reduction in Hydrotreated Gas Oil

N. Marchal, S. Kasztelan and S. Mignard
Institut Français du pétrole
BP 311, 92506 Rueil-Malmaison, France

ABSTRACT

Deep desulfurization and dearomatization of gas oil can be obtained in a two-stage low pressure process. In the first stage deep HDS is achieved using a hydrotreating catalyst. In a second stage aromatic reduction is obtained using a hydrogenation catalyst. The use of either a hydrotreating catalyst or a sulfur resistant noble metal catalyst for the second stage aromatics reduction has been considered in this work. Catalytic tests have been performed on a deeply desulfurized light cycle oil obtained from the first stage hydrotreatment. For both types of catalysts the effects of feed sulfur content, catalyst stability and process parameters have been investigated.

INTRODUCTION

Aromatics and sulfur in diesel fuels will be severely limited in the near future in Europe, Japan and the USA. High aromatics content in diesel fuel is known to lower the fuel quality as aromatics have low cetane index (1). In addition, aromatics in diesel fuel have been reported to contribute significantly to undesired emissions in exhaust gases from diesel engines (2,3). Diesel fuel contains also higher amount of sulfur than gasoline.

Because of the health hazards associated with these emissions, environmental regulations governing the composition of diesel fuels are being enacted in Europe and the USA (4,5). A 500 ppm sulfur content in diesel fuel is scheduled in the USA this year and the same limit is planed for Europe in 1996. Total aromatic content of diesel fuel will be limited to 35 vol % or a minimum cetane index of 40 in the USA. In some countries or states, very low aromatic content will be enforced such as a 10 vol % limit in California.

The total aromatic content of diesel feedstocks can vary considerably from 20 to 40 wt % in straight run middle distillates to 40-70 wt % in cracked distillates and as much as 90 wt % in catalytically cracked Light Cycle Oil (LCO). Thus moderate to severe hydrogenation will be required depending on feedstocks.

Existing gas oil hydrotreaters designed to reduce sulfur levels are capable of reducing aromatics content only marginally (6,7,8). However the 0.05 wt % sulfur target leads refiners to operate hydrotreaters at more severe conditions (lower LHSV) than usual, which will contribute to reduce the total aromatic content and increase the cetane index (9).

High pressure hydrotreating is one approach to increase aromatic saturation (6-13). However when very low total aromatic contents (10 wt % or less) are required or highly aromatic feedstocks are processed, very high hydrogen pressure will be needed (12). Few refiners have such high pressure hydrotreaters available and such high pressure capacity is extremely expensive to built (10, 11).

Another approach to reach low aromatic content in diesel fuels involves the use in combination of a hydrotreating stage using a deep HDS catalyst to reach the 500 ppm sulfur level or less and a hydrogenation stage using a sulfur resistant hydrogenation catalyst, both stages operating under moderate pressure. In between the two stages, the desulfurized diesel oil is stripped from H_2S, NH_3 and light hydrocarbons.

Two-stage low pressure processes have been largely emphasized in recent years (1,6,9,12-15). Such processes offer flexibility, lower capital cost than a high pressure single-stage process and deeper aromatic reduction.

Operating conditions of the second stage will depend on the degree of impurity removal in the first stage. We have investigated in this work the catalytic properties of a NiMo hydrotreating catalyst and of a noble metal based sulfur resistant catalyst in the conditions of the second stage of a two-stage process such as the IFP process (12). The sulfur tolerance of both catalysts have been evaluated as well as their stability.

EXPERIMENTAL

A deeply Desulfurized Light Cycle Oil (DLCO) feed has been prepared by hydrotreating a pure LCO. The main properties of both feedstocks are summarized in Table 1. The DLCO has a very low sulfur and a high aromatic content. The sulfur content of this feed has been increased artificially up to 1000 ppm S by adding dibenzothiophene (DBT). This procedure allows to evaluate the effect of the amount of sulfur for the same aromatic composition of the feed.

The DBT additivated DLCO has been processed in a pilot plant under second-stage conditions using a commercial NiMo hydrotreating catalyst (HR348 from PROCATALYSE) or a commercial noble metal based sulfur resistant catalyst designed for deep aromatic reduction (LD402 from PROCATALYSE).

Aromatics content in feed and products were measured by mass spectrometry and ^{13}C NMR. These methods give the total content of aromatic carbon in wt %. The FIA method (ASTM D1319) prescribed by the EPA as a standard method for determining total aromatic content in diesel fuel gives the total aromatic content in vol % and does not give the distribution of mono-, di- and tri-aromatics in the feed. The different methods employed to measure total aromatic content have been analyzed by others (11,13,16) and correlations have been developed by several workers (7,16) between total aromatics content determined by FIA and other methods. In this work mass spectrometry has been used for total aromatic content and for mono-, di-, tri-aromatic distribution determination.

RESULTS AND DISCUSSION

Preparation of the DLCO feed by first stage hydrotreating.

The deeply Desulfurized Light Cycle Oil (DLCO) feed has been obtained by hydrotreating the pure LCO feed using a deep HDS catalyst (HR348 from PROCATALYSE). The very low sulfur content of DLCO (Table 1) has been obtained in rather mild hydrotreating conditions i.e. less than 40 bar

(590 psi) hydrogen pressure,   SOR temperature of 360°C (680°F) and LHSV= 0.9h^{-1}. Although deep HDS and hydrodenitrogenation have been reached, the total aromatic content of the DLCO feed remains high, demonstrating the limitations of low pressure single-stage hydrotreating. However the distribution among mono-, di-, tri-aromatics has changed considerably (Table1). Whilst tri$^+$-aromatics and diaromatics have decreased substantially, monoaromatics have increased considerably. This trend results from the hydrogenation sequence :

$$\text{Triaromatics} \longrightarrow \text{Diaromatics} \longrightarrow \text{Monoaromatics}$$

with accumulation of monoaromatics which are the most difficult compounds to hydrogenate.

Table 1 : Properties of Feedstocks

	LCO	DLCO
S wt ppm	21800	18
N wt ppm	625	5
Spec. Grav.	0.945	0.896
D86 (°C)		
IBP	184	177
10%	246	225
50%	283	263
90%	339	316
EP	372	359
Aromatics (MS, wt %)		
Total	80	69
Mono-	17	48
Di-	39	19
Tri-	11	2
Aro S	13	0
Aromatic Carbon (^{13}C)	55	36

Sulfur resistance of hydrogenation catalysts.

It is well known that catalysts based on noble metals such as platinum on alumina catalysts exhibit excellent hydrogenation activity but are poisoned by very small amounts of sulfur compounds in the feedstocks (17). Thus a severe hydrodesulfurization step is necessary to reduce the sulfur content to a level that does not affect the performance of the Pt/Al$_2$O$_3$ catalyst. Levels of sulfur content tolerable by such catalysts are often set at one ppm or less (17). Over a Pt/alumina catalyst, low space velocities were required to substancially reduce aromatics content of a feed containing 1.5 ppm S according to ref. 13. In ref. 17 a Pt/silica-alumina catalyst has been reported to be able to process feed containing up to 10 ppm S whereas noble metal based zeolite catalysts are said to be able to reduce aromatics content of feed containing up to 600 ppm S (14).

The sulfur resistance of a noble metal can be improved by using zeolites as a support as reported by (17, 19). Dispersion of the noble metal is also another factor that has been reported to favor sulfur resistance (17).

The use of such types of catalysts allows either to limit the severity of the first-stage operating conditions or to reach very deep aromatic reduction if the feed contains low sulfur content.

To illustrate the sulfur resistance of the commercial noble metal based catalyst used in this work, catalytic tests have been performed on the DLCO feed in which various amounts of DBT have been added in order to vary the amount of sulfur without modifying the total aromatics content and the mono-, di,- tri-aromatics distribution. These tests have also been performed using the NiMo catalyst in the same operating conditions.

Each catalyst has been found to have a different response to the sulfur content of the feed which has been increased artificially up to 1000 ppm S by adding dibenzothiophene (Figure 1). Whilst the activity of the NiMo catalyst increased when the sulfur content of the feed increased, the activity of the noble metal LD402 catalyst decreased. Although it is known that at high partial pressure, H$_2$S poisons the hydrogenation sites of NiMo hydrotreating catalysts, it is also known that H$_2$S is necessary to keep the active phase in the sulfided state (12). Thus, the increase of the hydrogenation activity with the increase of sulfur corresponds to the maintenance of the sulfided state of the catalyst.

The noble metal based LD402 catalyst was found more active than the NiMo catalyst up to 1000 ppm sulfur in the feed. The LD402 catalyst has a remarquable sulfur resistance. In addition, for the mild operating conditions used, high aromatic reduction can be achieved. Clearly the lower the sulfur content of the feed the higher the hydrogenation activity of the LD402 catalyst. At less than 20 ppm S almost 100 % aromatic reduction can be reached at 60 bar (880 psi) total pressure.

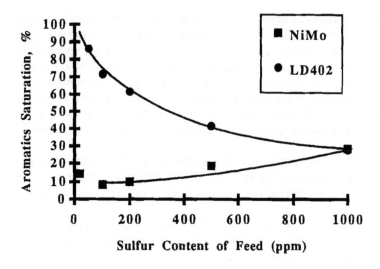

Figure 1 : Sulfur resistance of NiMo and LD402 Catalysts

It was found that the degree of aromatics saturation increased with increase in reactor temperature up to a maximum level. The temperature at which the maximum conversion was obtained was a function of pressure and space velocity. Further increase in temperature resulted in a decrease in aromatics saturation. These results are illustrated in figure 2 in which the degree of aromatics saturation is plotted against temperature for two different hydrogen/oil ratio. The existence of a maximum conversion at a given pressure indicates a limitation imposed by thermodynamic equilibrium. This is supported by the mass spectrometry aromatics analysis reported in figure 3 which shows that, at about 320°C (608°F), the equilibrium between monoaromatics and naphthenes is established because of the increase of the monoaromatics content when the reaction temperature increases.

The effect of the increase of the H_2/oil ratio is to increase the hydrogenation activity, an effect which has also been reported in ref. 13. Then in figure 2 for an H_2/oil ratio of 900 l/l, the thermodynamic equilibrium is reached at a slightly lower reaction temperature.

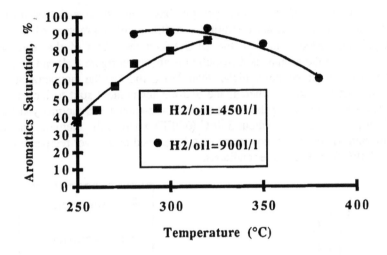

Figure 2 : Performance of LD402 on DLCO Feed

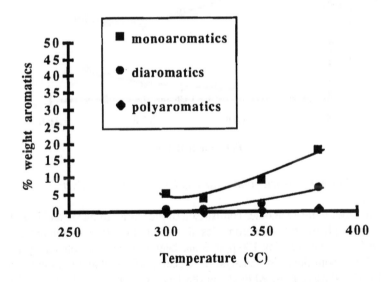

Figure 3 : Distribution of aromatics for LD402

The effect of reaction temperature has been also investigated for the NiMo catalyst for various operating conditions. At a pressure of 60 bar (880 psi) and 1000 ppm S in the DLCO feed the effect of the reaction temperature on the aromatics distribution is shown in figure 4. It can be seen that at reaction temperatures higher than 350°C (680°F) the monoaromatics content start to increase whereas the diaromatics content start to increase at 320°C (608°F) indicating that the monoaromatic-naphthenes thermodynamic equilibrium is reached at about 350°C (680°F) whereas the thermodynamic equilibrium diaromatics-monoaromatics is reached at a lower temperature at about 320°C (608°F) in these conditions.

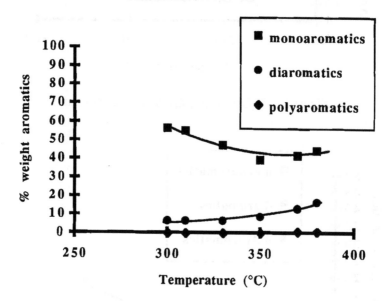

Figure 4 : Distribution of aromatics for NiMo Catalyst

In order to make another comparison of the ability of NiMo and LD402 catalysts to deeply hydrogenate aromatics, the effect of the reaction temperature on the DLCO feed containing 100 ppm S has been reported for both catalyst in figure 5. Tests with the NiMo catalyst have been performed at a total pressure of 120 bar (1760 psi) compared to 60 bar (880 psi) for the LD402 catalyst. In figure 5, it can be seen that doubling the pressure is not enough for the NiMo catalyst to reach the aromatics saturation level of the noble metal based LD402 catalyst.

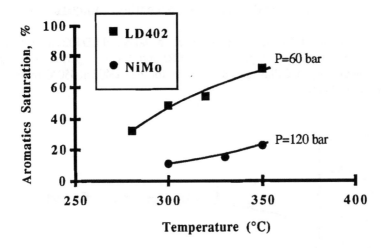

Figure 5 : *Performance of LD402 and NiMo on DLCO Feed*

Hydrodesulfurization by second-stage catalysts.

Besides the aromatics reduction brought about by the second stage catalyst, some hydrodesulfurization and hydrodenitrogenation are also performed. The sulfur contents of the hydrogenated DLCO are presented in table 2 for several reaction temperatures and sulfur contents of the feed. The hydrogenated DLCO feed has not only a reduced aromatic content but has also very small amounts of sulfur left and a nitrogen content below the detection limit of 0.5 ppm. Table 2 shows that whatever the amount of sulfur added as DBT to the DLCO feed, both the NiMo and LD402 catalyst efficiently desulfurize the feed.

Table 2: HDS Performances of NiMo and LD402 (2 ppm : limit of detection)

Catalyst	T(°C)	S feed	S product
NiMo	300	200	3
	350	200	<2
	300	515	4
	350	515	<2
	350	1000	<2
LD402	250	15	<2
	320	100	<2
	280	200	5
	350	1000	3

Stability of noble metal based hydrogenation catalysts.

Both the NiMo catalyst and the noble metal based LD402 catalyst have shown excellent stability in presence of 1000 ppm S over a period of several hundred hours. Figure 6 illustrates the stability for the LD402 at 350°C (680°F) and 60 bar (880 psi).

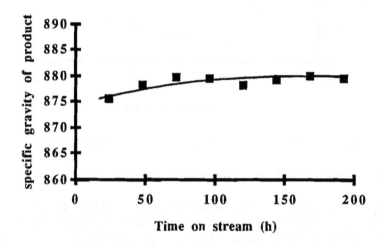

Figure 6 : Stability of LD402 on feed DLCO + 1000 ppm S

Another test of the LD402 stability has been performed by exposure of the catalyst to successive spikes of 1000 ppm S additivated DLCO while operating with the DLCO (15 ppm S). The aromatics saturation obtained, in the steady state, are presented in table 3. The reversibility of the LD402 catalyst to spikes of high sulfur content feed appears remarkable.

Table 3 : Stability of LD402 (Ptot = 60 bar)

Temperature (°C)	S feed (ppm)	Aromatics saturation (%)
320	15	81
350	1000	28
320	15	80

CONCLUSION

The refining industry is asking for a low cost approach to both deep desulfurization and dearomatization while maximizing the use of existing equipment. These constraints have led us to focus on low pressure operation. While desulfurization is relatively easy to accomplish, meeting future regulations regarding aromatics content will be more difficult especially for diesel feedstocks with high aromatic contents. One approach is two-stage low pressure processes using deep HDS catalyst in the first stage and a sulfur resistant hydrogenation catalyst in the second stage. The results reported in this work show clearly that the hydrogenation activity of the sulfur resistant LD402 catalyst will be higher on low sulfur content feed. Then the two-stage process will be best optimized for deep aromatic reduction by performing HDS as deep as possible in the first stage hydrotreating. By varying processing conditions and catalysts, the combination of deep HDS and aromatic saturation can achieve products with less than 500 ppm sulfur and less than 10 % vol aromatics.

REFERENCES

1. Suchanek, A.J., "Catalytic Routes to Low Aromatics Diesel Look Promising", Oil and Gas J., p. 109, May 7, 1990.

2. Ullman, T.L., "Investigation of the Effects of Fuel Composition on Heavy-Duty Diesel Engine Emissions", SAE Paper 892072, International Fuels and Lubricants Meeting and Exposition, Baltimore, Maryland, September 26-28, 1989.

3. Miller, C., Weaver, C.S. and Johnson, W., "Diesel Fuel Quality Effects on Emissions, Durability and Performance", Final Report EPA Contract 68-01-65443, Sept. 30, 1985.

4. Crow, P. and Williams, B., "US Refiners Facing Squeeze Under New Federal State Air Quality Rules", Oil and Gas J., p. 15, January 23, 1989.

5. Parkinson, G., "Stricter Standards for California Diesel Fuel", Chem. Eng., Vol. 96, p.42, January 23,1989.

6. Johnson, A.D., "Study Shows Marginal Cetane Gains from Hydrotreating", Oil and Gas J., p. 79, May 30, 1983.

7. Asim, M.Y. and Yoes, J.R., "Confronting New Challenges in Distillates Hydrotreating", paper AM-87-59, NPRA, March 1987.

8. McCulloch, D.C., Edgar, M.D. and Pistorious, J.T., "High Severity Diesel Hydrotreating", paper AM-87-58, NPRA, March 1987.

9. Lee, S.L. and de Wind, M., "Cetane Improvement of Diesel Fuels with Single and Dual Stage Hydrotreating", Preprints, Div. Petr. Chem., ACS, p 718, vol. 37, 1992.

10. Nash, R.M., "Process Conditions and Catalyst for Low-aromatics Diesel Studied", Oil and Gas J., p.47, May 29, 1989.

11. Asim, M.Y., Keyworth, D.A., Zoller, J.R., Plantenga, F.L. and Lee, S.L., "Hydrotreating for Ultra-low Aromatics in Distillates", paper AM-90-10, NPRA, March 1990.

12. Billon, A., Hennico, A., Peries, J.P. and Kressmann, S., "IFP Deep Hydrodesulfurization and Aromatics Hydrogenation of Straight Run and Pyrolysis Middle Distillates", paper AM-91-38, NPRA, March 1991.

13. Cooper, B.H., Stanislaus, A. and Hannerup, P.N., "Diesel Aromatics Saturation : A Comparative Study of Four Catalyst Systems", Preprint, Fuel Chem. Div., ACS, p 41., Vol. 37, 1992.

14. Milam, S.N., Winquist, B.H.C., Murray, B.D. and Del Paggio, A.A., "Reduction of Aromatics in Diesel Fuel. A Low Pressure Two Stage Approach", Presented at the Spring Meeting, AIChE, April 1991.

15. Eastwood, D., Tong, C. and Yen, C.S., "Integrated Hydrotreater/Aromatic Saturator Process for the Production of Oils with Reduced Aromatics and Polynuclear Aromatics", paper AM-92-63, NPRA, March 1992.

16. Somogyvari, A.F., Oballa, M.C. and Herrera, P.S., "Industrial Catalysts for Aromatics Reduction in Gas Oil", Preprints, Div. Petr. Chem., ACS, p 1878, Vol. 36, 1991.

17. Barbier, J., Lamy-Pitara, F., Marecot, P., Boitiaux, J.P., Cosyns, J., and Verna, F., "Role of Sulfur in Catalytic Hydrogenation Reactions", Advances in Catalysis, 37, 279, 1990.

18. Suchanek, A.J., "Reduction of aromatics in Diesel Fuel", paper AM-90-21, NPRA, March 1990.

19. Gallezot, P., "The State and Catalytic Properties of Platinum and Palladium in Faujasite-type Zeolites", Catal. Rev. Sci. Eng. 20, 121, 1979.

20. Le Page, J.F., "Applied Heterogeneous Catalysis", Technip, Paris, 1987.

18   Unionfining: Technical Case Studies

Tuan A. Nguyen and Milan Skripek

Unocal Fred L. Hartley Research Center *
Brea, California 92621

ABSTRACT

Hydrotreating improves the quality of FCC feeds by reducing sulfur, nitrogen, metals, asphaltenes, and polynuclear aromatic content. Four case studies presented in this paper show the benefits of hydrotreating FCC feeds: higher conversion and gasoline yield, better quality products, and lower SO_x emissions.

INTRODUCTION

FCC feed components derived from heavy crudes are high in sulfur, nitrogen, polynuclear aromatics, asphaltenes, carbon residue and metals -- all contributing to low yields and poor quality FCC products. VGO UNIONFINING is a Unocal process for hydrotreating wide boiling range straight run and cracked gas oils (a process flow diagram is shown in Figure 1). This process achieves high levels of HDM, HDN, HDS and polynuclear aromatics saturation at relatively low to moderate pressures with specifically designed catalyst systems. VGO UNIONFINING upgrades low quality

---

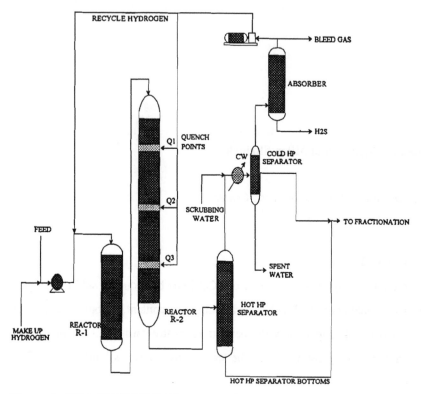

Figure 1: *VGO UNIONFINING Process Flow Diagram*

feedstocks to high quality feedstocks, producing high yields of environmentally acceptable FCC products and significantly reducing stack emissions.

This paper presents four pilot plant case studies to show the effects of hydrotreating severity on FCC yields and product qualities.

OBJECTIVES OF FCC FEED HYDROTREATING

The major objectives of FCC feed pretreatment are summarized below. The degree to which each of these objectives is carried out varies with the needs and constraints of the refiner.

1) **Hydrodesulfurization (HDS)**
2) **Hydrodenitrogenation (HDN)**
3) **Hydrodemetallation (HDM)**
4) **Aromatic Saturation (HDA)**
5) **Reduce Conradson Carbon Residue**
6) **Produce high quality diesel directly from the hydrotreater**

Sulfur removal to improve environmental performance

The increase in sulfur content of U.S. crude slate is troubling because it opposes the objective of providing lower sulfur gasoline and diesel (see Figure 2).[1] Hydrodesulfurization (HDS) of FCC feed can greatly reduce the sulfur content of gasoline, light cycle oil, and decant oil. The reduction of sulfur in FCC gasoline is especially important because this sulfur can account for more than 90% of the total sulfur in the gasoline pool.[2] Sulfur in gasoline is known to cause the temporary deactivation of catalytic converters and thereby increase exhaust emissions. Figure 3 shows sulfur contents for typical gasoline blending components, and Figure 4 shows the average FCC naphtha contribution to the 1989 U.S. gasoline pool.

HDS of FCC feed also lowers the amount of sulfur in spent catalyst coke, which results in the reduction of SOx emissions from the regenerator. However, deep desulfurization is needed to reduce the coke sulfur to the near zero level required by legislation in some areas, such as California. For a given percentage of desulfurization of the feed, the lighter FCC products will show a greater percentage reduction in sulfur, while the heavier products (coke and decant oil) will show a lesser percentage reduction.

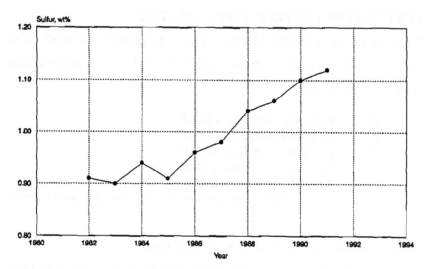

Figure 2: *Sulfur Content of U.S. Crude Slate*

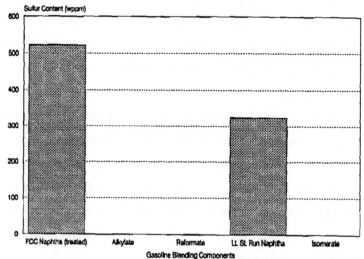

Figure 3: *Sulfur Contents for Typical Gasoline Blending Components*

For most feedstocks, low to moderate severity hydrotreating can be used to achieve a high level of desulfurization (≥90%).

Nitrogen removal to reduce coke propensity

Nitrogen compounds poison the active acid sites in the FCC catalyst.[3] Most catalysts include both zeolite and matrix acidity. Since the zeolite acidity is generally stronger, these sites tend to poison first, reducing the ratio of zeolite/matrix cracking. In this way, nitrogen compounds increase the coke making propensity of the catalyst and reduce the yield of valuable products.

Nitrogen removal by hydrotreating is much harder than sulfur removal. Moderate to high severity hydrotreating is generally required to achieve a high denitrogenation level (≥70%).

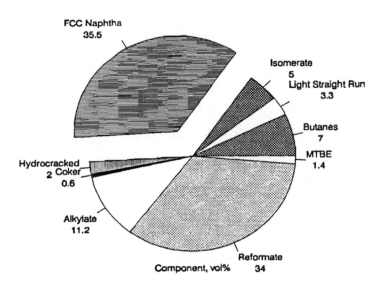

Figure 4: *Average 1989 U.S. Gasoline Pool*

Metals removal to permit better utilization of feedstocks

Nickel promotes dehydrogenation reactions which lead to increased coke and gas yields. Vanadium promotes the collapse and irreversible destruction of zeolites.[4] Although passivation technologies are available to mitigate some of these effects, the nickel and vanadium contents often dictate the allowable endpoint of the VGO's used for FCC feed. By quantitatively removing contaminant metals, FCC feed pretreating permits the refiner to take a deeper cut in the vacuum tower and this incremental VGO can be diverted to the FCC instead of the coker. Figure 5 shows the increased in nickel and vanadium levels on equilibrium catalysts.[5]

Complete metals removal can be achieved with low to moderate severity hydrotreating.

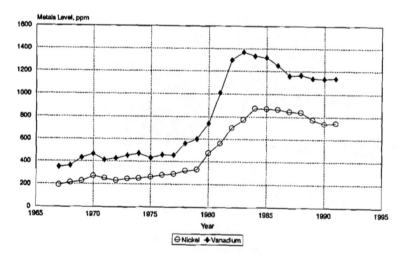

Figure 5:   *Contaminant Metals on Equilibrium FCC Catalysts (in U.S. & Canada)*

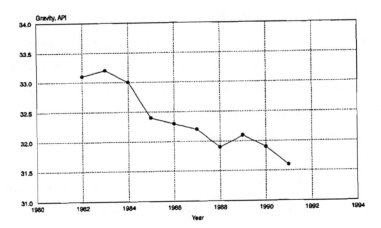

Figure 6: *API Gravity of U.S. Crude Slate*

Saturating aromatics to increase liquid yields

The decline of API gravity of U.S. crude slate over the last 10 years (see Figure 6) indicates that U.S. refiners are continuing to process more refractory feedstocks which contain higher levels of polynuclear aromatic compounds (PNA's). PNA's do not crack in an FCC unit but instead undergo condensation reactions to form coke.[6] Hydrotreating can saturate PNA's into easily cracked naphthenes and paraffins. Mono-aromatics, another product of PNA hydrogenation, resist cracking in an FCC unit, but lose side chains and end up providing high-octane components in the FCC gasoline.

The degree of aromatic saturation is highly dependent on hydrogen partial pressure, catalyst type, and the space velocity of the hydrotreater. The hydrogen content of the FCC feed governs the potential yields of valuable products. For each FCC unit, with its unique feedstock and operating conditions, there is an optimum hydrogen uptake for maximizing liquid products.

Removing Conradson Carbon Residue to reduce delta coke

Conradson Carbon Residue (CCR) is a known coke precursor. Eliminating CCR reduces delta coke (carbon on spent catalyst minus carbon on regenerated catalyst), resulting in lower regenerator temperatures and greater catalyst circulation rates for a given heat load. The reduction in delta coke and regenerator temperature can provide many benefits, particularly for units which operate near a coke limit.

Conradson Carbon Residue in most VGO feedstocks can be reduced below detectable limits at relatively low hydrotreating severity.

FCC FEED PRETREATING CASE STUDIES WITH VARIOUS GAS OIL FEEDSTOCKS

Four case studies are presented in the following sections. The first case study involves hydrotreating a commercial FCC feedstock at three levels of severity and examining the effects that hydrotreating has on FCC yields and product properties. The second case study examines the flexibility to swing the hydrotreater/FCC operation from the production of maximum gasoline to the production of maximum diesel. The third case study examines the possibility of reducing capital investment by hydrotreating only the most refractory FCC feed components. The fourth case study looks at using excess FCC hydrotreating capacity to upgrade the quality of a light cycle oil (LCO) stream.

Description of Pilot Plant Equipment

The experimental studies described in this paper were performed at Unocal's Fred L. Hartley Research Center in Brea, California. The hydrotreating runs were carried out in a multi-purpose isothermal pilot plant. This plant is capable of processing 5-30 liters/day, and it can be used to simulate any of Unocal's hydroprocessing technologies. Unocal's experience has shown that this unit matches commercial performance. It is particularly well suited for determining hydrogen consumption, product yields and properties, catalyst life, and operating conditions.

The FCC data were obtained on Unocal's circulating FCC pilot plant. This unit, which can process 25-50 liters/day, has been designed to closely simulate commercial operations. Test runs have shown that the product yields and selectivities, including coke make, obtained from this pilot plant duplicate those obtained from commercial units.

Case Study #1: FCC's Yields and Product Properties at Three Levels of Hydrotreating Severity.

In this study, a high-nitrogen, high-sulfur feedstock was blended from West Coast USA derived gas oils. Approximately 50% of the feed consisted of cracked stocks from commercial coking and visbreaking units. The nitrogen compounds in this feedstock were very refractory.

The feed was hydrotreated at three severity levels. Each of these severity levels was designed to provide a minimum 2.5 year catalyst life. The table below summarizes the feedstock properties for the untreated feedstock and

Table 1 *Properties of Untreated and Treated VGO's- Case 1*

Relative HT Severity-->	Untreated	Low	Moderate	High
Hydrotreating Conditions				
Pressure	—	Base	Base * 2.0	Base * 2.3
LHSV	—	Base	Base * 0.5	Base * 0.5
Temperature, ^{0}C	—	Base	Base + 15	Base + 30
Properties of Untreated and Treated VGO's				
Gravity, kg/m^3	937	908	893	870
Sulfur, wppm	27,000.	3,880.	250.	54.
Nitrogen, wppm	3,850.	3,060.	380.	23.
Con. Carbon, wt%	0.24	<0.10	<0.10	<0.10
Hydrogen, wt%	11.77	12.47	12.97	13.47
H$_2$ consumption, m^3/m^3 of feed	—	62	118	220

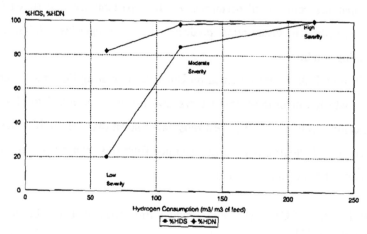

Figure 7: *% HDS and % HDN as a function of Hydrogen Consumption (Case 1)*

for the hydrotreated product from each of the three severity levels. The hydrotreater effluent was fractionated before FCC processing to remove the small amount of naphtha produced in the hydrotreater.

Percent HDS and HDN as a function of hydrogen consumption

At low severity, about 85% HDS and 20% HDN were obtained. At the moderate severity, the HDS and HDN had increased to 99% and 90%, respectively. At the high severity level, both HDS and HDN were above 99%. Most of the nitrogen and sulfur were already removed at the moderate severity level. In moving to the high severity level, aromatics saturation (HDA) increased substantially, as evidenced by the near doubling of the hydrogen consumption (see Figure 7). This study was conducted without recycle gas (H_2S) scrubbing.

Substantial Improvements are obtained from FCC feed pretreatment

Table 2 shows the product yields obtained when each of the four feeds in this study were processed in our FCC pilot plant. Compared to the base case (untreated feed), low severity hydrotreating increased the gasoline yield and conversion by 10.3 wt% and 13.3 wt%, respectively. The dry gas yield dropped substantially; this can be very important to units operating near a wet gas compressor limit.

Moving from the low to the moderate severity level boosted the gasoline yield and conversion another 6.6 wt% and 10.0 wt%, respectively. In addition, the coke make dropped substantially as a result of the HDN and HDA accomplished at the moderate hydrotreating severity. FCC feed preheat must be increased to compensate for lower coke make in order to maintain the unit's heat balance.

Moving from the moderate to the high severity level boosted the conversion level another 7.2 wt%, while the gasoline yield remained about the same. The extra conversion came almost entirely in the LPG fraction. Although the yields are expressed here in wt%, an examination of the vol% yields shows substantial swell in the products owing to the higher hydrogen content of the highly treated feed.

Table 2 *FCC Yields from Treated and Untreated VGO's- Case 1*

Relative HT Severity-->	Untreated	Low	Moderate	High
C_2-minus gases, wt%	4.9	3.1	2.3	2.5
Total LPG, wt%	8.7	13.5	19.6	26.9
Gasoline, wt%	39.0	49.3	55.9	56.1
Light Cycle Oil, wt%	27.5	21.5	15.2	9.6
Decant Oil, wt%	13.8	6.5	2.8	1.2
Coke, wt%	6.1	6.1	4.2	3.7
**Conversion, wt%	58.7	72.0	82.0	89.2

** Conversion is defined as weight percentage of FCC feed changed to C_2-minus gases, LPG, gasoline, and coke.

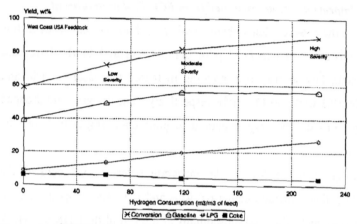

Figure 8: *FCC Yields as a Function of Hydrotreating Severity (Case 1)*

Figure 8 plots key yields as a function of hydrogen consumption. The graph clearly shows that the gasoline selectivity peaks at the moderate pretreating severity. Beyond that point, the lower PNA and higher hydrogen contents of the more severely treated feed induce lower coke make and higher conversion. However, these are accompanied by significant over-cracking of gasoline to LPG.

The choice of optimum operation is dependent on local product values and unit constraints. For example, if alkylation capacity is available or if maximum octane-barrels are desired, the high severity hydrotreating option shown here would allow the refiner to produce the highest yield of FCC gasoline, high-octane alkylate, and MTBE feed. The FCC hydrotreater design should be tailored to the needs of an individual refiner.

Case Study #2: Flexible Hydrotreating/FCC Operation for Maximum Gasoline & Maximum Diesel Modes

In this study, the base feedstock was a mixture of gas oils derived from Mid-West USA and Canadian crudes which are similar to Arabian and Soviet crudes. Coker gas oil accounted for about 20% of the feed. Two hydrotreating severity levels were examined, representing low and moderate severity levels. This feed was not nearly as refractory as the West Coast USA feed previously described and there was no need to investigate a high severity hydrotreating operation.

Since it is often desirable to respond rapidly to seasonal changes in the marketplace, one of the primary goals of this project was to demonstrate the flexibility to swing the hydrotreater/FCC operation from the production of maximum gasoline to the production of maximum diesel.

Definition of Gasoline and Diesel Modes of Operation

We examined the operation of the hydrotreater/FCC complex in both maximum gasoline and maximum diesel modes. In the maximum gasoline operation, the hydrotreater effluent was fractionated to remove only the small quantity of naphtha produced. The remaining material was used as FCC feed. In the maximum diesel operation, both the naphtha and diesel were removed. This diesel stream was of high quality and met key diesel specifications. The

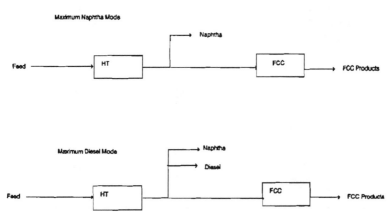

Figure 9: *Modes of Hydrotreater/FCC Operation (Case 2)*

block flow diagrams in Figure 9 illustrate these two modes of operation.

Yields and Diesel Properties from the Hydrotreater

The yields obtained from the hydrotreating operations are given in Table 3. The low and moderate severities were run at the same reactor temperature and space velocity, but the hydrogen partial pressure was about 40% higher in the moderate severity case. Note that the operating pressure had a negligible effect on the net conversion during hydrotreating. Each severity level was designed for a two year catalyst life.

The properties of the diesel fuel fractionated from the hydrotreater effluent during the maximum diesel operation met all key specifications and

Table 3 *Hydrotreater Product Yields and Properties- Case 2*

	Low Severity	Moderate Severity
Operating Conditions		
H_2 Partial Pressure	Base	Base*1.4
LHSV Base	Base	
Temperature, ^oC	Base	Base
H_2 consumption, m^3/m^3 of feed	85	136
C_1 - C_3 , wt%	0.15	0.23
Butanes, wt%	0.10	0.16
Naphtha, wt%	5.41	7.04
Diesel, wt%	37.1[**]	36.9[**]
Vacuum gas oil, wt%	55.6	54.4
Total FCC Feed, wt%ff		
max naphtha mode	92.7	91.3
max diesel mode	55.6	54.4
Diesel Properties		
Gravity, kg/m^3	871	865
Sulfur, ppm	50.0	10.0
Cetane index	44.7	45.4
Cetane #, engine	43.9	44.8
Pour Point, ^oC	-12	-15

[**]The hydrotreater feed contained about 25 wt% of diesel boiling range material, which was mostly coker distillate.

Table 4 *Properties of Untreated and Treated FCC Feeds- Case 2*

HT Severity———>	Untreated	Low Severity		Moderate Severity	
	Max.	Max.	Max.	Max.	Max.
Operating Mode——>	Naphtha	Naphtha	Diesel	Naphtha	Diesel
API gravity	23.6	28.1	27.0	30.3	29.2
Sulfur, wppm	17500	100	178	33	53
Nitrogen, wppm	1050	140	185	8	10
R.I. @ 67°C	1.4921	1.4759	1.4802	1.4650	1.4692
VABP, °C	416	376	433	376	437
UOP K index	11.75	11.96	12.10	12.12	12.27

had a relatively high cetane number.

FCC Feed Properties

The properties of the untreated and treated FCC feeds for both the maximum gasoline and maximum diesel cases are shown in Table 4. This base feedstock was much less refractory than the California gas oils described earlier. Note that even at low severity, over 99% of the sulfur and 86% of the nitrogen had been removed. At moderate severity, the sulfur and nitrogen removal were both over 99%. In addition, substantial PNA saturation took place, as evidenced by the improvements in refractive index and gravity.

FCC Product Yields

The FCC yields from the treated and untreated feeds are given in Table 5. For the purposes of this comparison, we have chosen to compare the yields at the peak of the gasoline/conversion curve, before the onset of over-cracking. Of course, this point would not necessarily represent the economic optimum. Downstream product handling, alkylation capacity, and incremental product values would determine the optimum FCC operation for each refiner.

All the hydrotreated feeds gave about a 25% decrease in dry gas yield, largely due to the virtual elimination of H2S. As mentioned earlier, this could be of considerable benefit for a unit which is limited by compressor capacity. Compared to the untreated feed, low severity pretreating improved the gasoline yield and conversion by 8.3 wt% and 9.9 wt%, respectively. As expected, the coke make dropped due to the reduced feed nitrogen and aromatic contents. Moderate severity pretreating further reduced the coke make and increased both the gasoline yield and conversion.

The FCC gasoline selectivity, as measured by the ratio of gasoline/conversion, was slightly better in the maximum diesel operation than in the maximum gasoline operation. Although this was expected due to the relatively low molecular weight of the diesel portion of the feed, the magnitude of the difference was much less than commonly believed. Calculation of yield vectors shows that the diesel fraction was an excellent FCC feed.

In this case study, the hydrotreated diesel fraction proved to be very effective as a "swing component" of the FCC feed. It should be noted that when operating the complex in the maximum diesel mode, it may be necessary to obtain a source of alternate VGO to maintain a sufficient FCCU throughput.

Table 5 *FCC Yields from Treated & Untreated FCC Feeds- Case 2*

HT Severity ——————→	Untreated	Low Severity		Moderate Severity	
	Max.	Max.	Max.	Max.	Max.
Operating Mode ——→	Naphtha	Naphtha	Diesel	Naphtha	Diesel
Dry Gas, wt% of feed	4.0	3.0	2.9	3.1	3.2
LPG, wt%	14.0	17.7	17.3	20.9	20.3
Gasoline, wt%	46.2	54.5	56.0	56.4	57.9
LCO, wt%	20.0	14.7	11.2	12.3	8.9
Decant Oil, wt%	9.7	5.1	7.9	3.6	6.0
Coke, wt%	6.0	5.0	4.6	3.7	3.7
**Conversion, wt%	70.3	80.2	80.9	84.1	85.1
C3+ liquids, wt%	90.0	92.0	92.5	93.2	93.1

** Conversion is defined as weight percentage of FCC feed changed to C2-minus gases, LPG, gasoline, and coke.

In the low severity hydrotreating case, the coke make listed for the diesel operation is lower than for the gasoline operation. Since the feed in the diesel case is somewhat easier to crack (owing to its higher molecular weight and VABP), the onset of over-cracking occurred at a slightly lower catalyst to oil ratio.

Overall Yields from Hydrotreater/FCC Complex

Summarized in the Table 6 are the overall yields from the hydrotreater/FCC complex, given on the basis of fresh feed to the hydrotreater (the untreated feed yields are given on the basis of feed to the FCC).

Swing Potential Between Maximum Gasoline and Maximum Diesel Modes

Figure 10 graphically depicts the swing potential between gasoline and diesel operations for the low severity hydrotreating case. The overall diesel yield (assuming the FCC LCO can be blended into the diesel pool) increased from

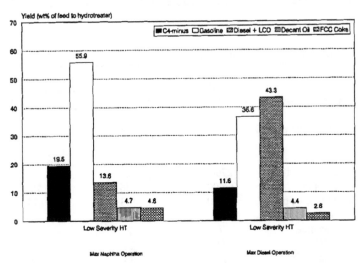

Figure 10: *Swing Potential between Operating Modes (Case 2 - Overall Yields from Hydrotreater/FCC Complex)*

13.6 wt% in the gasoline operation to 43.3 wt% in the diesel operation. To achieve this 30.3 wt% improvement in diesel yield, the gasoline yield dropped by 19.3 wt% (from 55.9 wt% to 36.6 wt%) and the C4-minus yield was reduced by about 8 wt%.

Coke yield was also lower in the diesel case because both the FCC coke make and the FCC throughput were reduced. As noted earlier, in order to "swing" the complex from gasoline to diesel operation, it may be necessary to supplement the FCC feed because many FCC units cannot handle such large turndown ratios (about 40% less feed) and maintain smooth operation.

Sulfur Distribution in FCC products

An additional benefit of FCC feed pretreating is that the FCC products have substantially lower sulfur contents. Table 7 shows the sulfur content of the FCC gasoline, light cycle oil, and decant oil. Even with low severity hydrotreating, the light cycle oil had less than 0.05 wt% sulfur. Although it has a low cetane index (about 15), the reduced sulfur level may allow the LCO

Table 7 *Distribution of Sulfur in FCC Products- Case 2*

HT Severity——> Operating Mode——>	Untreated Max. Naphtha	Low Severity Max. Naphtha	Moderate Severity Max. Naphtha
SOx emission, g/kg of feed	4.235	0.109	0.037
Sulfur content			
Gasoline, wppm	2700	6	<3
LCO, wppm	27400	146	70
Decant Oil, wppm	33700	443	156
Wt% of Feed Sulfur			
Gasoline	7	2	<1
LCO	31	21	26
Decant Oil	19	23	17
Coke	12	54	56
H2S	31	0	0

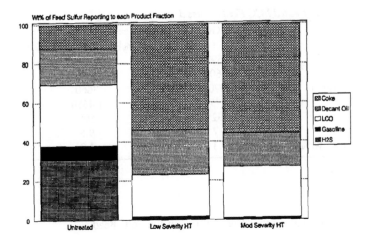

Figure 11: *Sulfur Distribution in FCC Products (Case 2)*

to be blended in the refinery diesel pool.

Hydrotreating the FCC feed dramatically changes the sulfur distribution in the FCC products (please see Figure 11). For example, in the case of the untreated feed, 31% of the FCC feed sulfur was produced in the lightest product, H_2S, while 12% of the feed sulfur was produced in the heaviest product, coke (and was burned in the regenerator to form SOx). With pretreated feedstocks, the FCC product sulfur, although much lower in absolute terms, tends to proportionally shift away from the lighter products and toward the heavier products. For example, the percentage of feed sulfur produced in the gasoline dropped from 7% in the untreated case to less than 1% in the moderately treated case. At the same time, the proportion of coke sulfur (SOx) increased from 12% to 56%. This nearly five-fold increase explains why deep desulfurization is needed to adequately control FCC SOx emissions. As a rule, hydrotreating preferentially removes sulfur compounds that crack to form H_2S; more refractory sulfur compounds not removed during hydrotreating tend to

Table 8 *Properties of Untreated FCC Feeds- Case 3*

	AGO	HCGO	HVGO	Feed to Hydrotreater 80% HCGO/20% HVGO
Gravity, kg/m^3	885	959	954	959
Sulfur, wppm	14,900	25,600	21,600	24,700
Nitrogen, wppm	890	4070	1990	3660
R.I. @ 67°C	1.4835	1.4759	1.4802	1.4650
EP, $^\circ$C	514	592	614	595
Ni+V, wppm	0.7	3.7	10.5	9.5
Con. Carbon, wt%	0.0	1.3	1.2	1.3

crack to form coke.[7, 8]

Case Study #3: Hydrotreating the "Bad Actors" in the FCC Feed

In this case study, only the most refractory FCC feed components were hydrotreated in order to minimize capital investment. The feedstocks used for this study included a high quality atmospheric gas oil (AGO), moderate quality heavy vacuum gas oil (HVGO), and poor quality heavy coker gas oil (HCGO). The HVGO and HCGO were derived from a South American crude. The "bad actor" feed to hydrotreated were a blend of 80% HCGO and 20% HVGO. Unlike the previous two case studies, yield data presented for this study will be in volume percentage of FCC feed rather than weight percentage.

Table 9 *Properties of Untreated and Treated "Bad Actor" FCC Feed- Case 3*

Relative HT Severity-->	Untreated	Low	Moderate
Hydrotreating Conditions			
Pressure	---	Base	Base*1.4
LHSV	---	Base	Base
Temperature, $^\circ$C	---	Base	Base-3
Gravity, kg/m^3	959	924	916
Sulfur, wppm	24700	2420	670
Nitrogen, wppm	3660	2010	865
Hydrogen, wt%	11.18	12.19	12.50
Ni+V, wppm	9.5	<0.1	<0.1
Con. Carbon, wt%	1.3	<0.1	<0.1

Two levels of hydrotreating were evaluated. The low severity case was carried out at operating conditions chosen to provide about 90% HDS and 50% HDN. Operating conditions for the moderate severity case were chosen to provide about 99% HDS and 75% HDN. Table 9 shows a summary of the hydrotreated product from each of the two severity levels. The hydrotreater effluent was fractionated before FCC processing to remove any diesel-minus product.

A matrix of seven FCC runs was developed to span a wide range of hydrotreater/FCC options: from the lowest capital investment case in which 0% of the FCC feed was hydrotreated (Case 3A) to the moderate capital investment cases in which only the most refractory feed components were hydrotreated to the highest capital investment case in which 100% of the FCC feed was hydrotreated at moderate severity (Case 3G).

Case	AGO	HVGO	HCGO	Low HT	Moderate HT
3A	20%	40%	40%	0%	0%
3B	35	30	0	35	0
3C	35	30	0	0	35
3D	30	0	0	70	0
3E	30	0	0	0	70
3F	0	0	0	100	0
3G	0	0	0	0	100

Treating a portion of the FCC feed substantially improves FCC yields structure

Product yields and product properties at 4.6 wt% coke make are summarized in Table 10. Compared to the untreated feed (Case 3A), treating 35% of the FCC charge at low severity increased the conversion by 12.0 vol%. By treating 35% of the feed at moderate severity (Case 3C), the conversion and gasoline yield increased another 3.1 vol% and 1.0 vol%, respectively. The moderate severity hydrotreated feed was of high quality and when its proportion was increased from 35% to 70% (Case 3C versus Case 3E), the conversion and gasoline yield increased by 3.1 vol% and 3.9 vol%, respectively.

An "apparent" anomaly emerges in comparing Case 3E to Case 3G. The conversion and product selectivities for Case 3E (where the moderate severity hydrotreated component accounted for 70% of the FCC feed) were similar to those for Case 3G (where the moderate severity hydrotreated stock accounted for 100% of the FCC feed). This anomaly can be explained by the

Table 10 *FCC Yields and Product Properties for Various- Case 3*

Case -->	3A	3B	3C	3D	3E	3F	3G
FCC Feed Properties							
Gravity, kg/m³	947	928	924	928	913	924	916
Sulfur, wt%	2.13	1.19	1.16	0.51	0.44	0.20	0.07
Nitrogen, wt%	0.253	0.150	0.139	0.147	0.088	0.201	0.088
Hydrogen, wt%	11.65	12.14	12.22	12.38	12.65	12.19	12.54
Ni+V, wppm	5.2	3.1	3.2	0.2	0.2	<0.1	<0.1
Con. Carbon, wt%	1.0	0.4	0.3	0.2	0.1	<0.1	<0.1
FCC Yields							
Dry Gas, FOE%	6.7	6.1	5.8	4.8	4.9	5.1	4.7
LPG, vol%	17.5	24.5	26.5	23.5	27.7	22.2	29.4
Gasoline, vol%	39.6	50.0	51.0	50.5	54.9	49.9	55.1
LCO, vol%	26.3	22.4	20.9	23.0	19.4	21.2	17.3
Decant Oil, vol%	17.8	9.7	8.1	9.5	6.6	12.3	7.9
Coke, wt%	4.6	4.6	4.6	4.6	4.6	4.6	4.6
Conversion, vol%	55.9	67.9	71.0	67.4	74.1	66.5	74.8

fact that 30% of the untreated feed in Case 3E is high quality AGO, while 100% of the treated feed in Case 3G (while still of high quality) is derived from a refractory blend of 80% HCGO/ 20% HVGO. This "apparent" anomaly illustrates a very important point: significant improvement in FCC yields and operations can often be achieved by hydrotreating only the worst FCC feed components at moderate severity. However, it should be also noted that in environmentally sensitive areas it may be necessary to hydrotreat the entire FCC feed to comply with stack emissions and limitations on sulfur contents in gasoline and diesel.

     Figures 12 and 13 and show sulfur and nitrogen contents in the products as a function of sulfur and nitrogen contents in the feed.

Case Study #4: Using excess FCC hydrotreating capacity to upgrade LCO quality

Excess FCC hydrotreating capacity can be used to upgrade the quality of light cycle oil. In this study, a blend consisting of LCO, HVGO, and HCGO was hydrotreated at high severity conditions to achieve >99% HDS and >98%

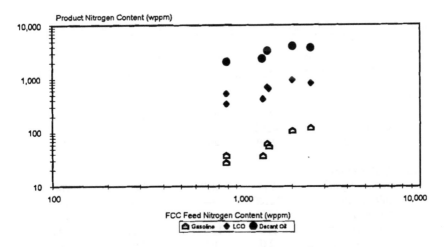

Figure 12: *Nitrogen Contents in FCC Products (Case 3)*

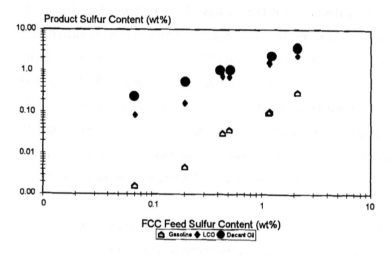

Figure 13: *Sulfur Contents in FCC Products (Case 3)*

HDN. Properties for the untreated LCO and properties for the diesel product obtained from the hydrotreater are given in Table 11. About 30% of the diesel yield from the hydrotreater effluent can be attributed to the LCO component in the hydrotreater feed.

Table 11 *Properties of Untreated and Treated "Bad Actor" FCC Feed- Case 3*

Feed to the Hydrotreater-->		Gas Oils	Gas Oils+LCO	Gas Oils+LCO
Relative HT Severity-->	Untreated	Base	Base + 6°C	Base + 22°C
Diesel Properties				
Gravity, kg/m^3	946	884	892	873
Sulfur	4540	---	36	12
Cetane Index	20.9	45.3	41.8	44.1
Gas Oil Properties				
Gravity, kg/m^3	934	897	895	874
Sulfur, wppm	22100	---	174	11
Nitrogen, wppm	1660	135	100	0.1
Hydrogen, wt%	11.9	13.11	13.17	13.57
Ni+V, wppm	2.0	0.0	0.0	0.0
Con. Carbon, wt%	0.6	<0.1	<0.1	<0.1

The untreated LCO would make a very poor blending component for the diesel pool because of its low cetane index and high sulfur content of 4540 wppm (the 1990 amendment to the Clean Air Act will limit the sulfur content of diesel fuel to less than 0.05 wt%). By combining this LCO stream with the gas oils going to the FCC hydrotreater, a good quality diesel stream can be obtained from the hydrotreater effluent; this diesel stream has a sulfur content of less than 40 wppm and a cetane index that is greater than 41.

INTERPLAY OF FCC FEED HYDROTREATING PROCESS VARIABLES

The FCC feed hydrotreating data presented in this paper represent a small fraction of the FCC hydrotreating data base Unocal has accumulated over the last three decades. This data base includes a wide range of atmospheric and vacuum distillates (cracked and straight run), many types of catalysts and varied process objectives. These data were utilized to develop generalized kinetics and correlations, which are used to optimize the designs and operations of commercial units and also to help guide the catalyst and process development efforts.

HDA, HDN, and HDS kinetics are generalized in the following equation.

$$\text{HDA/HDN/HDS} = \text{function}[(P_{H2})(P_{H2S})(P_{NH3})(\text{LHSV})(\text{Rx temp})(\text{Feed})(\text{Catalyst Type})]$$

HDS increases with temperature, H_2 partial pressure, and catalyst volume. HDS is strongly inhibited by H_2S, and scrubbing H_2S from recycle gas is frequently justified when desulfurizing high sulfur FCC stocks. HDS follows 1.5-2.0 reaction order kinetics as indicated by the asymptotic relationship between HDS and catalyst volume (Figure 14).

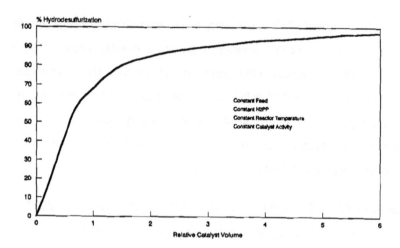

Figure 14: *% HDS vs. Catalyst Voume Required*

Nitrogen removal from FCC feedstocks also increases with temperature, catalyst volume, and H_2 partial pressure. NH_3 strongly inhibits HDN reactions, and H_2S has been found to enhance HDN activity in certain operations.

Aromatic saturation is strongly dependent on H_2 partial pressure, catalyst volume, and temperature (up to equilibrium limitations); H_2S and NH_3 can have an inhibiting effect on aromatic saturations.

The order of difficulty in removing FCC feed contaminants are listed as follows: aromatics, nitrogen, sulfur, heavy metals. In other words, metals (Ni, V) are easiest and aromatics are most difficult to remove from the FCC feeds.

Catalyst deactivation kinetics

Quantitative knowledge of the HDS, HDN, and HDA kinetics is essential in optimizing design and operation of FCC hydrotreaters. Equally important is the knowledge of the rate of catalyst deactivation.

The rate of catalyst deactivation increases with feed rate (space velocity), SOR temperature, the extent of heteroatom removal, feed contaminants (CCR, asphaltenes, metals etc.), and feed molecular weight. Catalyst life increases significantly with H_2 partial pressure and with catalysts designed to tolerate contaminants.

Figure 15 illustrates the effect of % HDS on the rate of catalyst deactivation. At constant H_2 partial pressure, space velocity, and feed, the rate of catalyst deactivation can be six times higher at 95% than at 60% HDS. Similar relationships apply to HDN and HDA reactions.

CONCLUSIONS

Upgrading the FCC feedstocks by hydrotreating will become more prevalent in the coming years as environmental constraints become more stringent and refinery crude slates become heavier and more difficult to process into environmentally acceptable products.

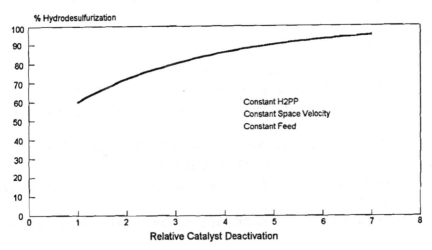

Figure 15: *Effect of % HDS on Catalyst Deactivation*

The data presented in this paper demonstrate that cracked and straight run vacuum distillates derived from a wide range of crudes can be upgraded to high quality FCC feedstocks by a commercially proven Unocal hydroprocessing technology, VGO UNIONFINING.

Unocal, with its extensive data base, large stable of proven catalysts, modern pilot plant facility, and generalized correlations has the capability to optimize designs of hydrotreaters for a wide range of feedstocks and process objectives. Unocal has licensed nineteen FCC pretreaters worldwide.

Optimum VGO UNIONFINING severity is governed by feedstock and process/product objectives. In environmentally sensitive areas it may be necessary to hydrotreat the entire FCC feed to comply with stack emissions and limitations on sulfur in gasoline and diesel. Often, significant improvement in FCC yields and operations can be achieved by hydrotreating only the worst FCC feed components at moderate severities.

Excess FCC hydrotreating capacity can also be used to upgrade a low cetane, high sulfur light cycle oil stream into a high quality diesel blending component.

REFERENCES

1. Swain, E. J., Oil & Gas Journal, March 1, 1993, pp. 62-64.
2. Unzelman, G. H., Fuel Reformulation, July/August 1992, pp. 28-31.
3. Fu, C. M., and A. M. Schaffer, Ind. Eng. Chem. Prod. Res. Dev., Vol. 24, 198 68-75.
4. Occelli, M. L., Catal. Rev.- Sci. Eng., Vol. 33 (3&4), 1991, pp.241-280.
5. Davison Catalagram, Number 83, 1992, pp.36-40.
6. Wojciechowski, B. W., and A. Corma, Catalytic Cracking, Marcel Dekker, York, 1986.
7. Wollaston, E. G., W. L. Forsythe, and I. A. Vasalos, Oil & Gas Journal, Aug 1971, pp. 64-69.
8. Huling, G. P., and J. D. McKinney, Oil & Gas Journal, May 19, 1975, pp.73-7

19   Effect of H$_2$S on the Functionalities of a Co/Mo/Al$_2$O$_3$ Hydrotreating Catalyst

J. van Gestel, J. Leglise and J.-C. Duchet.

Laboratoire Catalyse et Spectrochimie,
URA CNRS 04.414, ISMRA,
Université de CAEN.
14050 CAEN Cedex, FRANCE.
Fax: (+33) 31 45 28 77.

Address correspondence to J.-C. Duchet.

Abstract

The addition of various amounts of H$_2$S during the reaction of thiophene, cyclohexene and 2,6-dimethylaniline model compounds was used to assess the functionalities of an industrial CoMo/Al$_2$O$_3$ hydrotreating catalyst operating at 4 MPa total pressure. The catalyst was sulfided either by H$_2$S/H$_2$ or DMDS(dimethyldisulfide)/H$_2$. Thiophene and cyclohexene were converted simultaneously at 280°C, while 2,6-dimethylaniline was studied separately at 300°C. The alkylaniline reaction followed three parallel routes which were influenced only up to 1 mol% added H$_2$S: hydrogenation of the benzenic ring and direct C-N bond rupture to m-xylene were strongly inhibited, whereas acidic disproportionation was enhanced. H$_2$S affected thiophene and cyclohexene differently: in the triangular reaction network of thiophene, hydrogenation of thiophene to the intermediate thiolane appeared promoted, and the C-S hydrogenolysis steps were depressed; cyclohexene hydrogenated to cyclohexane much more slowly with H$_2$S, and reacted significantly to cyclohexanethiol. Thiolane and cyclohexanethiol acted as strong inhibitors. Sulfiding the catalyst with DMDS/H$_2$ instead of H$_2$S/H$_2$ yielded more acidic and C-S hydrogenolysing properties but less hydrogenating capability. The effect of H$_2$S on the functionalities of the catalyst is discussed in terms of active species arising from dissociated H$_2$ and H$_2$S.

Keywords: hydrotreating catalyst, cobalt-molybdenum/alumina, hydrodesulfurization, hydrogenation, hydrodenitrogenation, thiophene, cyclohexene, 2,6-dimethylaniline, dimethyldisulfide, hydrogen sulfide.

Introduction

Hydrotreating of petroleum fractions proceeds mainly through hydrogenation and hydrogenolysis steps for removal of sulfur and nitrogen. Depending on the feedstock to treat, high levels of hydrogen sulfide and ammonia can be produced. H_2S is by far the most abundant (Schulz et al., 1986; Kasztelan et al., 1991) and its influence on various reactions has been the subject of numerous studies.

Generally, hydrogenation steps are slightly inhibited or unaffected by H_2S (Gultekin et al., 1984; Ramachandran and Massoth, 1981), whereas activity for sulfur removal is strongly depressed at high levels of H_2S (Vrinat, 1983; Radomyski et al., 1988). Conversely, H_2S promotes the C-N bond breaking of saturated amines (Ho, 1988; Massoth et al., 1990; Perot, 1991; Hadjiloizou et al., 1992), although inhibition was found in the HDN of carbazole (Nagai et al., 1988); cracking and isomerization are also enhanced by H_2S (Ramachandran et al., 1981; Yang and Satterfield, 1984). Apparent discrepancies between the results on the effect of H_2S can be ascribed to different structure of the organic reactant, and to various experimental conditions, such as hydrogen pressure and temperature.

Recently, we reported how H_2S effected the conversion of 2,6-dimethylaniline (DMA) at 300°C and 4 MPa (Van Gestel et al., 1992a). Hydrogenation and direct C-N hydrogenolysis were depressed only by addition of H_2S amounting to less than 1 mol% and remained strictly constant beyond, whereas disproportionation increased rapidly at low H_2S levels.

In the present work, we examine the effect of H_2S on the conversion of a complex feed containing thiophene, cyclohexene and m-xylene over a CoMo/Al$_2$O$_3$ catalyst at 280°C and 4 MPa. Results are gathered with the previous ones for 2,6-dimethylaniline. The aim is to draw an extended picture of the influence of H_2S on the functionalities of the sulfided hydrotreating catalyst, i.e. C-S and C-N bond breakages, hydrogenation and acidity. In addition, we studied the effect of sulfiding either by H_2S/H_2 or $CH_3SSCH_3(DMDS)/H_2$.

Experimental

Catalyst

The CoMo/Al$_2$O$_3$ catalyst, from AKZO (KF 742-1.3 Q), had worked in a hydrotreating plant and had been regenerated. Elemental analysis (CNRS, Vernaison) gave 2.8 wt.% Co, 9.3 wt.% Mo, 0.5 wt.% C, and 0.6 wt.% S as sulfate species. The catalyst was free from vanadium and nickel.

Sulfidation and catalytic test

Experiments were performed in a high pressure flow reactor operating at 4 MPa total pressure. The reactor was loaded with 30 mg of oxide catalyst (grain size 0.6-1 mm) diluted with SiC. The samples were in-situ sulfided either with a mixture of H$_2$S/H$_2$ (10 mol% H$_2$S) produced by decomposition of dimethyldisulfide (DMDS) in a prereactor, or with DMDS/H$_2$ (5 mol% DMDS); DMDS was introduced by means of a liquid flow regulator.

The procedure started with pressurization of the system to 4 MPa; then the temperature of the reactor was raised by 3°C min^{-1} up to 350°C under the sulfiding mixture (total flow 40 cm^{3}min^{-1}) and maintained for 12 h. In the case of DMDS/H$_2$, the sulfidation started with 1 mol% DMDS up to 150°C to avoid liquid condensation, then with 5% at higher temperature.

At the end of the sulfidation period, the catalyst was cooled down to 280°C for the catalytic test. The liquid feed containing thiophene (T, 13 vol.%), cyclohexene (CHE, 7%), m-xylene (19%), n-heptane (51%) and 2-propanol (10%) was introduced with a metering pump at a dosing rate of 0.02 cm^{3}min^{-1}. 2-propanol prevented gumming of the pump head by cyclohexene. The feed was totally vaporized in the system under hydrogen flow, and the H$_2$S content was adjusted (0-5 mol%). After stabilization of the catalyst for 24 h, the space-time was changed, keeping constant the partial pressure of the organic reactants (thiophene, 14.6 kPa; cyclohexene, 6.3 kPa; m-xylene, 13.2 kPa; n-heptane 30.3 kPa; 2-propanol 12.0 kPa).

On-line analysis was carried out every 2 h with a GC equipped with a CPSIL-5B capillary column. Steady-state conversions were measured after 12 h on stream. The stability of the catalyst was checked over 2 weeks.

The reaction of 2,6-dimethylaniline (DMA) was studied in a second system following the same procedure. The conditions were 300°C, and 4 MPa. DMA (1.3 kPa) was diluted in a mixture of n-hexane and n-heptane. Details are reported elsewhere (Van Gestel et al., 1992a).

Results

Among the compounds of the complex feed, n-heptane and m-xylene did not react under the experimental conditions (280°C, 4 MPa). 2-propanol was almost totally converted into C_3 hydrocarbons and H_2O, even at low space-time. Therefore, only the simultaneous reactions of thiophene and cyclohexene were examined with respect to H_2S. The effect of H_2S on the reaction of 2,6-dimethylaniline was studied separately.

Thiophene reaction

The products identified from thiophene (T) conversion were n-butenes and n-butane (C4), and thiolane (tetrahydrothiophene, THT) which appeared as an intermediate. The reaction obeyed a triangular network (Scheme 1) (Joffre et al., 1991; Leglise et al., 1992):

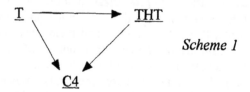

Scheme 1

Although the reaction of thiophene ran simultaneously with cyclohexene, separate experiments showed that thiophene was almost unaffected by cyclohexene.

The influence of H_2S (up to 4.7 mol%, 188 kPa) added to the feed on the conversion and on the yields of products at a given low thiophene space-time ($W/F_T = 6.7$ hgmol^{-1}) is shown in Figure 1 for the CoMo/Al$_2$O$_3$ catalyst sulfided with H_2S/H_2 or DMDS/H_2. The thiophene conversion sharply decreased at low H_2S content, then more slowly. The yield of THT went through a flat maximum, while the C4 production continuously decreased. The DMDS/H_2 sulfided catalyst converted more thiophene than the H_2S/H_2 counterpart without addition of H_2S; but gave almost the same conversion at high levels of H_2S. The amount of THT was lower after DMDS/H_2 sulfiding. The same behaviour was found for every space-time between 3 and 60 hgmol^{-1}.

The selectivity into THT at various levels of added H_2S is shown in Figure 2 for the H_2S/H_2 sulfided catalyst. The yield of THT clearly increases with the H_2S concentration. The THT selectivity for the DMDS/H_2 sulfided catalyst is reported on Figure 2 only for 1.3% added H_2S; it is much lower than after H_2S/H_2 sulfiding. From the THT yield measured at low conversion, it follows that formation of THT is half-order with respect to H_2S.

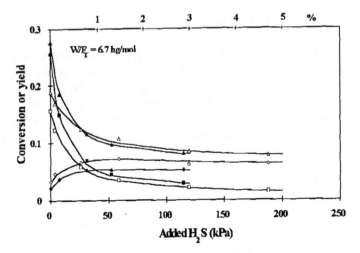

Figure 1: Thiophene reaction at 280°C and 4 MPa over sulfided CoMo/Al₂O₃ catalyst: effect of added H₂S on the conversion of thiophene (Δ, ▲) and on the yields of C₄ hydrocarbons (□, ■) and thiolane (◊, ◆). Open symbols: sulfidation with H₂S/H₂; filled symbols: sulfidation with DMDS/H₂.

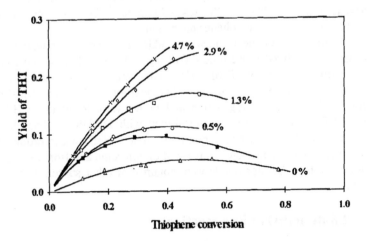

Figure 2: Thiophene reaction at 280°C and 4 MPa over sulfided CoMo/Al₂O₃ catalyst: effect of added H₂S (mol%) on the selectivity for thiolane (THT). Open symbols: catalyst sulfided with H₂S/H₂; filled symbols: 1.3 mol% added H₂S for the catalyst sulfided with DMDS/H₂.

Cyclohexene reaction

The major reaction of cyclohexene (CHE) was hydrogenation to cyclohexane (CHA). Cyclohexanethiol (CHS) was also formed in small quantities (< 6 mol%); it arised from addition of H_2S and matched thermodynamics at sufficient high space-time (Van Gestel et al., 1992b). Isomerization to methylcyclopentenes was negligible at 280°C. The corresponding reaction network (Scheme 2) was deduced from previous analysis (Leglise et al., 1992):

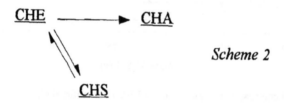

Scheme 2

The conversion of cyclohexene was strongly inhibited by the simultaneous thiophene reaction. Indeed, in a separate experiment, cyclohexene reacted twice as fast when the complex feed did not contain thiophene. Moreover, replacing thiophene by thiolane decreased the conversion of cyclohexene from 30 to 20% at $W/F_{CHE} = 70$ hgmol^{-1}. Therefore, cyclohexene is strongly sensitive to the thiophene conversion.

Figure 3 shows the influence of added H_2S on the conversion and yields measured at fixed space-time ($W/F_{CHE} = 16$ hgmol^{-1}) on the catalyst sulfided with H_2S/H_2 or $DMDS/H_2$. The two catalysts were hardly distinguished without added H_2S, but after addition of H_2S the $DMDS/H_2$ sulfiding yielded lower hydrogenation and higher thiol formation. CHS was equilibrated faster on the catalyst presulfided with $DMDS/H_2$. Cyclohexene conversion went through a slight minimum at low H_2S concentration. This curve resulted from a continuous decrease of cyclohexane and from an increase of cyclohexanethiol. Similar trends were obtained at other space-times (7-140 hgmol^{-1}).

2,6-dimethylaniline reaction

The conversion of DMA at 300°C and 4 MPa has already been reported on the CoMo/Al$_2$O$_3$ catalyst presulfided with H_2S/H_2 (Van Gestel et al., 1992a). In the present study, the catalyst sulfided by $DMDS/H_2$ was also examined.

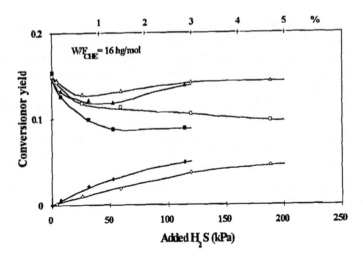

Figure 3: Cyclohexene reaction at 280°C and 4 MPa over sulfided CoMo/Al$_2$O$_3$ catalyst: Effect of added H$_2$S on the conversion of cyclohexene (Δ , ▲), on the yields of cyclohexane (□ , ■) and cyclohexanethiol (◊ , ◆). Open symbols: sulfidation with H$_2$S/H$_2$; filled symbols: sulfidation with DMDS/H$_2$.

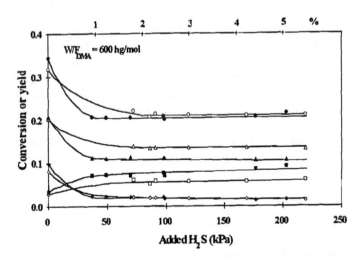

Figure 4: 2,6-dimethylaniline reaction at 300°C and 4 MPa over sulfided CoMo/Al$_2$O$_3$ catalyst: Effect of added H$_2$S on the conversion of 2,6-dimethylaniline (O , ●), on the yields of dimethylcyclohexanes + dimethylcyclohexenes (Δ , ▲), m-xylene (◊ , ◆), and mono- + trimethylaniline (□ , ■). Open symbols: sulfidation with H$_2$S/H$_2$; filled symbols: sulfidation with DMDS/H$_2$.

Three parallel reaction routes were identified : i) a major hydrogenation of the benzenic ring, yielding 1,3-dimethylcyclohexenes and 1,3-dimethylcyclohexanes; ii) a minor direct C-N bond rupture to m-xylene; iii) a significant disproportionation of DMA to 1-methylaniline and 2,4,6-trimethylaniline.

The influence of H_2S on each route is shown in Figure 4 in which the mole fraction of the products were lumped according to the above-specified routes. The general trends are identical for both modes of sulfiding. Hydrogenation and C-N hydrogenolysis behaved identically with H_2S: they were strongly inhibited only up to 1 mol% of added H_2S. This effect is relatively more pronounced for hydrogenolysis than for hydrogenation. Conversely, disproportionation was increasing with H_2S.

Discussion

The various reactions of the model compounds studied involve different functionalities of the sulfided $CoMo/Al_2O_3$ catalyst: i) hydrogenation of the aromatics, dimethylaniline and thiophene, and of cyclohexene; ii) C-heteroatom bond rupture, C-S from thiophene and thiolane, and C-N from dimethylaniline; iii) acidic disproportionation of the alkylaniline.

The various responses of each reaction path on H_2S will be analyzed in order to determine for which functionality H_2S is beneficial or detrimental. This would be helpful in understanding the behavior of the catalyst in real operation.

In this respect DMA was clearly distinguished from CHE and T: H_2S influences the former only in the range 0-1 mol% H_2S, while it constantly affects the other two reactants.

Effect of H_2S on the reactions

2,6-dimethylaniline

Results over the $DMDS/H_2$ sulfided catalyst complete previous data over the H_2S/H_2 sulfided catalyst (Van Gestel et al., 1992a). They follow the same trends (Fig. 4): H_2S depresses hydrogenation and direct C-N bond rupture only between 0 and 1 mol% H_2S. On the contrary, H_2S favors disproportionation.

As already stated by van Gestel et al. (1992a), the behaviour of the first two reactions cannot be attributed to competitive adsorption between DMA and other reactants, especially H2S. The data obtained on the H2S/H2 sulfided catalyst were rather interpreted in terms of mechanistic pathways, involving electrophilic and nucleophilic substitution on the aromatic ring with H^+ and H^- species respectively. Without H2S, the active species arise from heterolytic splitting of hydrogen. Moreover, we suggested that H2S blocks a fraction of these sites. However, because of the lower relative inhibition of hydrogenation compared to that of hydrogenolysis it is likely that part of the H2S is dissociated. The resulting H^+ species participates in hydrogenation, whereas SH⁻ is not active for direct C-N bond splitting.

The catalyst exhibits some activity for the third route yielding disproportionation. This reaction is related to acidity which is enhanced by dissociation of H2S (Hadjiloizou et al., 1992; Topsoe et al., 1989). The relative distribution of active species and acidity is illustrated in Scheme 3 without and for more than 1 mol% added H2S.

This scheme holds also for the DMDS/H2 sulfided catalyst. This catalyst was as efficient as the H2S/H2 sample for xylene formation from DMA. Since this reaction requires H⁻ species created by heterolytic dissociation of H2, we infer that H2 dissociates to the same extent on both catalysts (22%, Scheme 3). Consequently, the concentrations of H^+ arising from H2 are also equal on both samples.

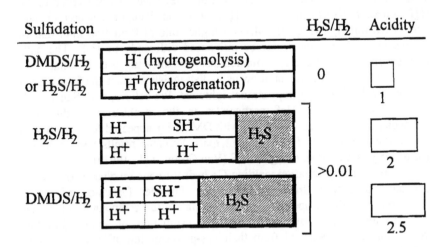

Scheme 3: Normalized distribution of active species and acidity over CoMo/Al₂O₃ catalyst sulfided by H2S/H2 or DMDS/H2, without addition of H2S and for more than 1 mol% of added H2S.

The difference between the two modes of sulfiding for hydrogenation, higher on the H_2S/H_2 sulfided catalyst, is attributed to a larger fraction of dissociated H_2S on the hydrogen sites (48% against 28%). Besides, the $DMDS/H_2$ sulfiding favors disproportionation after addition of H_2S which enhances acidity, presumably on separate sites.

Thiophene-cyclohexene

By contrast with DMA, thiophene and cyclohexene are continuously affected by H_2S. They were run simultaneously, but the results allow separate discussion of the thiophene reaction since it is almost unaffected by cyclohexene.

Thiophene

The triangular reaction Scheme involves hydrogenation and C-S hydrogenolysis steps.

Hydrogenation of thiophene to thiolane is unlikely to proceed directly. As already stated (Joffre et al., 1991; Leglise et al., 1992) it should involve a dearomatization step, yielding semi-hydrogenated compounds such as dihydrothiophenes (Schulz et al., 1986; Zdrazil and Kraus, 1986); these compounds are too reactive to be detected but they should be concerned in the THT formation. Kinetically, since THT is an intermediate, its increased amount with increasing H_2S (Fig. 2) reflects a higher rate of formation than that of disappearance. These rates may be influenced differently by H_2S. The half-order dependence of the ratio formation/disappearance suggests dissociation of H_2S but the rough data do not allow us to distinguish whether the formation is promoted by H_2S or the disappearance inhibited. We believe, with Sauer et al. (1989), that dissociated H_2S participates in the mechanism of thiophene hydrogenation. The catalyst sulfided with $DMDS/H_2$ appeared less effective in THT formation than its H_2S/H_2 counterpart. Previous results favor a higher hydrogenolysis rate when using DMDS (Van Gestel et al., 1992b).

Hydrodesulfurization yielding C_4 hydrocarbons involves two kinds of C-S bond breakages (from T and THT). The net formation of C_4 is continuously inhibited upon H_2S addition. But sulfur compounds other than H_2S are also competitive inhibitors. An estimation of their adsorption coefficients can be obtained from their gas-phase proton affinity by analogy with the effect of nitrogen compounds on thiophene (La Vopa et al., 1988) or dibenzothiophene (Nagai et al., 1986) reactions. Then H_2S is found a much weaker inhibitor than NH_3 in hydrogenation of propylbenzene (Gultekin et al., 1984) or p-cresol (Gevert et al., 1987). Indeed, the proton affinity of H_2S is much lower (172 kcal mol^{-1}) (Gasteiger and Hutchings, 1984) than that of ammonia (207 kcal mol^{-1}) (Aue et al., 1984).

Proton affinities for the sulfur compounds of interest are not available, but from a series of thiols and thioethers (Gasteiger and Hutchings, 1984), we estimate a proton affinity value for THT greater than or equal to that of ammonia. CHS would have a slightly lower proton affinity than THT. On the other hand, protonation of thiophene is not favoured on the heteroatom (159 kcal mol^{-1}) (Houriet et al., 1981). Values for other compounds such as water, 2-propanol, and m-xylene lie below 193 kcal mol^{-1} (Gasteiger and Hutchings, 1984; Chong and Franklin, 1972a). Low values are also reported for olefins (Chong and Franklin, 1972b).

Accordingly, the adsorption coefficients of the strongest inhibitors should follow the sequence THT > CHS >> H₂S > T. Thus, THT and, to a lesser extent CHS are mainly responsible for the decreasing conversion of thiophene. Since the inhibitors are produced by reaction with H₂S, the inhibition can be considered as an indirect effect of H₂S. The reaction of thiophene is very sensitive at low H₂S levels (< 1 mol%) where THT and CHS concentrations steeply increase.

The production of C₄ is greater on the DMDS/H₂ sulfided catalyst, especially at very low H₂S levels; that agrees with a lower formation of THT (Fig. 1). At high H₂S levels, the formation of CHS becomes substantial so that both sulfided catalysts are as much efficient for hydrodesulfurization.

Cyclohexene

Cyclohexene conversion shows a complex response upon H₂S addition (Fig. 3). To clarify the global effect, we consider first the effect on thiol formation then on hydrogenation to cyclohexane.

Thiol formation is an unavoidable reversible addition of H₂S onto the cycloolefine. Under our conditions, the thermodynamic equilibrium is reached more readily over the DMDS/H₂ catalyst than over the H₂S/H₂ catalyst. Therefore this reaction is catalytic. It probably involves acidic dissociated H₂S, and in this respect, the DMDS/H₂ sulfided catalyst shows the highest ability, in agreement with the increased yield of disproportionation products observed in the reaction of DMA.

Hydrogenation to cyclohexane was considerably lowered in the range 0-1 mol% H₂S. This contrasts with the scarce results in the literature concluding to no effect of H₂S on hexene hydrogenation at atmospheric pressure (Ramachandran and Massoth, 1981), and to a slight decrease in propylbenzene conversion under hydrogen pressure (Gultekin et al., 1984). In fact, our reaction is strongly inhibited by THT. In the simultaneous conversion of cyclohexene and thiophene, THT is always present in variable amounts and competes with cyclohexene on the hydrogenation sites. The thiol also contributes partly to the inhibition.

The modes of sulfiding are not distinguished without addition of H_2S (Fig. 3), but with added H_2S the conversion to cyclohexane is more depressed on the DMDS/H_2 catalyst. This is surprising because of the lower yield of THT on this type of catalyst.

Finally, m-xylene in the complex feed suffers from the presence of THT and CHS which hampers its hydrogenation.

Active species

The influence of H_2S on the functionalities of the sulfided CoMo/Al_2O_3 catalyst points to the important role of dissociated H_2 and H_2S, yielding adsorbed H^+, H^- and SH^- species. They act specifically depending on the reactant molecule.

Among the nucleophiles, H^- is the only species able to break the C-N bond directly from the aromatic aniline (Van Gestel et al., 1992a). SH^- is known to assist E2 or SN mechanism in the C-N bond cleavage of saturated amines (Ho, 1988; Perot, 1991). Such a role of SH^- may also be considered in C-S bond rupture of cyclic thioethers such as THT and dearomatized thiophene. As for nitrogen compounds, it would start with protonation of the heteroatom, which weakens the C-N or C-S bond. From the proton affinities, this should be more difficult for sulfur than for nitrogen compounds; protonation is then expected to be rate-limiting for the C-S bond rupture compared to the subsequent SH^- attack in the α or β position. This scheme can be considered for THT and dearomatized T, but cannot apply to the thiophene molecule since protonation is unlikely on the heteroatom (Houriet et al., 1981). Therefore direct extrusion of sulfur from the aromatic ring is unlikely.

On the other hand, H^+ appears to be involved in many steps of the reactions: disproportionation of 2,6-dimethylaniline, dearomatization of DMA and of thiophene, hydrogenation of and H_2S addition onto cyclohexene, protonation at the heteroatom before C-S bond breakage. However, these reactions require different strengths of protonic acidity. Protons arising from hydrogen or hydrogen sulfide after dissociation on the hydrogen sites are probably involved in dearomatization and hydrogenation (Scheme 3). Those needed for disproportionation are stronger and we propose that they are created on separate sites (referred as acidity in Scheme 3). They might also act in the cyclohexanethiol formation which is recognized as an acid-assisted reaction (March, 1985).

The formation of THT from dihydrothiophene seems to be promoted by H_2S although we should expect an inhibition as for cyclohexene hydrogenation. The reason is not clear, but could be related to a mechanism involving specific superficial SH groups (Sauer et al., 1989).

Performing the reaction with H2S, always present in hydrotreatment, has drastic consequences on the reactivity. H2S enhances the formation of THT and yields CHS, both acting as strong inhibitors. The reaction conditions should be adjusted to minimize their undesirable influence. Thus, increasing the temperature will lower the adsorption coefficients of the inhibitors, and also the concentration of THT since its hydrogenolysis is more activated than its formation (Van Gestel et al., 1992b).

From our results, the mode of sulfiding is pertinent to the final properties of the catalyst in the presence of H2S. Sulfiding with DMDS/H2 increases the amount of acidic sites (increased yields of thiol and disproportionation compounds). It also enhances C-S hydrogenolysis involving the nucleophiles H⁻ and SH⁻, but has no effect on direct aromatic C-N hydrogenolysis. The drawback is a lower capability for hydrogenation by H⁺ due to H2S adsorption on the hydrogen sites, as illustrated in Scheme 3. DMDS/H2 sulfiding appears preferable because it yields less THT intermediate.

Conclusion

As the main conclusion of this study, a significant test of a sulfide catalyst should be performed with a feed containing several model molecules, to reveal simultaneously the different functionalities of the catalyst .

Under such conditions, the influence of H2S is complex: by direct action, it enhances the formation of inhibitors, thiolane from thiophene and cyclohexanethiol from cyclohexene, and depresses the dissociation of hydrogen molecules. Consequently, the yields of C4 hydrocarbons from thiophene hydrodesulfurization and of cyclohexane from cyclohexene hydrogenation are continuously decreased upon H2S addition.

Without strong inhibitors, H2S diminishes hydrogenolysis and hydrogenation of 2,6-dimethylaniline only in the range 0-1 mol% H2S. The effect is negligible beyond.

All steps of the reactions studied are controlled by dissociated H2 and H2S species.

Acknowledgement

One of us, J.v.G., thanks Société Nationale ELF Aquitaine for financial support.

References

Aue, D.H., H.M. Webb, and M.T. Bowers, "A Thermodynamic Analysis of Solvatation Effects on the Basicities of Alkylamines. An Electrostatic Analysis of Substituent Effects," *J. Am. Chem. Soc.*, **98**, 318 (1976).

Chong, S.L., and J.L. Franklin, "Proton Affinities of Benzene, Toluene and the Xylenes," *J. Am. Chem. Soc.*, **94**, 6630 (1972a).

Chong, S.L., and J.L. Franklin, "Heats of Formation of Protonated Cyclopropane, Methylcyclopropane and Ethane," *J. Am. Chem. Soc.*, **94**, 6347 (1972b).

Gasteiger, J. ,and M.G. Hutchings, "Quantitative Models of Gas-Phase Proton Transfer Reactions Involving Alcohols, Ethers, and their Thio Analogues. Correlation Analyses Based on Residual Electronegativity and Effective Polarizability," *J. Am. Chem. Soc.*, **106**, 6489 (1984).

Gevert, B.S., J.E. Otterstedt, and F.E. Massoth, "Kinetics of the HDO of Methyl-substituted Phenols," *Appl. Catal.*, **31**, 119 (1987).

Gultekin, S., S.A. Ali, and C.N. Satterfield , "Effects of hydrogen Sulfide and Ammonia on Catalytic Hydrogenation of Propylbenzene," *Ind. Eng. Chem. Process Des. Dev.*, **23**, 179 (1984).

Hadjiloizou, G.C., J.B. Butt, and J.S. Dranoff, "Catalysis and Mechanism of Hydrodenitrogenation: the Piperidine Hydrogenolysis Reaction," *Ind. Eng. Chem. Res.*, **31**, 2503 (1992).

Ho, T.C., "Hydrodenitrogenation Catalysts," *Catal. Rev. Sci. Eng.*, **30**, 117 (1988).

Houriet, R., H. Schwarz, W. Zummack, J.G. Andrade, and P. Von Ragné Schleyer, "α- vs β-Protonation of Pyrrole, Furan, Thiophene, and Cyclopentadiene. Gas Phase Proton and Hydrogen Affinities. The Bishomocyclopropenyl Cation," *Nouv. J. Chim.*, **5**, 505 (1981).

Joffre, J., P. Geneste, J.B. Mensah, and C. Moreau, "An Attempt to Rationalize Hydroprocessing of Cyclic Heteroaromatics over Sulfided Catalysts through a Common Limit Step: Hydrodearomatization," *Bull. Soc. Chim. Belg.*, **100**, 865 (1991).

Kasztelan, S., T. des Courières, and M. Breysse, "Hydrodenitrogenation of Petroleum Distillates: Industrial Aspects," *Catal. Today*, **10**, 433 (1991).

La Vopa, V., and C.N. Satterfield, "Poisoning of Thiophene Hydrodesulfurization by Nitrogen Compounds," *J. Catal.*, **110**, 375 (1988).

Leglise, J., J. Van Gestel, and J.C. Duchet, "Alkyl Polysulfides as Presulfiding Agents of a CoMo/Al₂O₃ Catalyst: Effect on Catalytic Properties," *Phosphorus, Sulfur, and Silicon*, **74**, 479 (1992).

March., J, *Advanced Organic Chemistry: Reactions, Mechanism and Structure*, Wiley-Interscience, N.Y., 3rd Edition, Chapter 15 (1985).

Massoth, F.E., K. Balusami, and J. Shabtai, "Catalytic Functionalities of Supported Sulfides. VI . The Effect of H₂S Promotion on the Kinetics of Indole Hydrogenolysis," *J. Catal.*, **122**, 256 (1990).

Nagai, M., T. Masagunaga, and N. Hana-oka, "Hydrodenitrogenation of Carbazole on a Mo/Al₂O₃ Catalyst. Effect of Sulfiding and Sulfur Compounds," *Energy and Fuels*, **2**, 645 (1988).

Perot, G., "The Reactions Involved in Hydrodenitrogenation," *Catal. Today*, **10**, 447 (1991).

Portefaix, J.L., M. Cattenot, M. Guerriche, J. Thivolle-Cazat, and M. Breysse, "Conversion of Saturated Cyclic and Noncyclic Amines over a Sulphided NiMo/Al₂O₃ Catalyst: Mechanisms of Carbon-Nitrogen Bond Cleavage," *Catal. Today*, **10**, 473 (1991).

Radomyski, B., J. Iszczygiel, and J. Trawczynski, "Reaction of Thiophene with Hydrogen over CoMo/g-Al₂O₃ Catalysts. II The Kinetics of the Reaction," *Appl. Catal.*, **39**, 25 (1988).

Ramachandran, R., and F.E. Massoth, "The effect of H₂S on the Hydrogenation and Cracking of Hexene over a CoMo Catalyst," *J. Catal.*, **67**, 248 (1981).

Sauer, N.N., E.J. Markel, G.L. Schrader, and R.J. Angelici, "Studies on the Mechanism of Thiophene Hydrodesulfurization: Conversion of 2,3- and 2,5-Dihydrothiophene and Model Organometallic Compounds," *J. Catal.*, **117**, 295 (1989).

Schulz, H., M. Schon and N.M. Rahman, "Hydrogenative Dinitrogenation of Model Compounds as Related to the Refining of Liquid Fuels," *Stud. Surf. Sci. cat.*, **27**, 201 (1986).

Topsoe, N.Y., H. Topsoe, and F.E. Massoth, "Evidence of Bronsted Acidity on Sulfided Promoted and Unpromoted Mo/Al₂O₃ Catalysts," *J. Catal.*, **119**, 252 (1989).

Van Gestel, J., J. Leglise, and J.C. Duchet, "Effect of Hydrogen Sulfide on the Reaction of 2,6-Dimethylaniline over Sulfided Hydrotreating Catalysts," *Appl. Catal., A: General*, **92**, 143 (1992a).

Van Gestel, J., J. Leglise, and J.C. Duchet, "Catalytic Properties of a CoMo/Al$_2$O$_3$ Catalyst Presulfided with Alkyl Polysulfides: Comparison with Conventional Sulfiding," *J. Catal.*, in press (1992b).

Vrinat, M.L., " The Kinetics of the Hydrodesulfurization Process - A Review," *Appl. Catal.*, **6**, 137 (1983).

Yang, S.H., and C.N. Satterfield, "Catalytic Hydrodenitrogenation of Quinoline in a Trickle-Bed Reactor. Effect of Hydrogen Sulfide," *Ind. Eng. Chem. Process Des. Dev.*, **23**, 20 (1984).

Zdrazil, M., and M. Kraus, "Effect of Catalyst Composition on Reaction Networks in Hydrodesulfurization," *Stud. Surf. Sci. Catal.*, **27**, 257 (1986).

20 New Hydroprocessing Catalysts Prepared from Molecular Complexes

Teh C. Ho

Corporate Research Laboratories
Exxon Research and Engineering Company
Annandale, NJ   08801

Hydroprocessing is an integral part of fuels and lubes refining (1,2). Most refinery streams are hydroprocessed. Depending on the severity of the conditions, the reactions taking place in a hydroprocessing reactor include hydrodesulfurization (HDS), hydrodenitrogenation (HDN), hydrodeoxygenation, hydrogenation of aromatics (or HDA, standing for hydrodearomatization), and hydrocracking. These reactions collectively reduce air pollution, protect catalysts in downstream processing, and improve product quality and stability. In the face of mounting public concerns over environment and a dwindling supply of high-quality crudes, there is no doubt that hydroprocessing will become increasingly important in years to come. Tremendous efforts have expended on the development of improved or new catalysts.

Current commercial hydroprocessing catalysts are transition metal sulfides (TMS) based on Group VIII and VI metals. They are prepared by dispersing MoO_3 (typically 12-25 wt %) and a promoter metal oxide, either CoO or NiO (typically 3-6 wt %), on γ-Al_2O_3 or SiO_2-modified Al_2O_3. This is followed by sulfiding with a sulfur-bearing stream such as H_2S at high temperatures. The thus formed MoS_2 crystallites are the backbone of the working catalysts. MoS_2 has a structure built of closely-packed layers of sulfur atoms stacked to create triagonal prismatic inter-stices which are occupied by Mo atoms. The active sites are widely believed to be sulfur anion vacancies associated with exposed Mo cations whose structure and electronic properties can be strongly affected by the neighboring promoter metals. These sites occur naturally at the edges of MoS_2 crystallites. Several models of such sites have been proposed to account for the activity of commercial catalysts (3,4). Among them are the contact synergy model, the pseudo-intercalation model, and the CoMoS phase model. While there are significant differences among them, all models point to the importance of having the promoter metal ions intimately associated with MoS_2. The scanning Auger study of Chianelli et al. (5) have shown the enrichment of Co along the MoS_2 edge planes. This enrichment tends to cause MoS_2 to grow thicker.

Based on the above, a potentially fruitful approach to new catalysts would be to molecularly

incorporate promoter metals into the structure of MoS_2
edge planes. Unfortunately, the details of MoS_2 edge
structure at atomic resolution are not yet available.
The edge can terminate in many different ways while
still maintaining local electrical neutrality.   A
structure showing some of the possible surface species
is given by Stiefel et al. (6). From this hypothe-
sized structure, one can identify several moieties
whose structures are reminiscent of those of homo- and
heteronuclear TMS complexes and clusters. Thus, as a
first step, it would seem reasonable to exploit the
use of heterometallic metal sulfur complexes as
hydroprocessing catalyst precursors. Controlled
synthesis of such complexes may well result in active
hydroprocessing catalysts.

Accordingly, in our laboratory we have developed
several families of new catalysts along this line.
Among them are the catalysts prepared from the following chemistries:

I.   Thermal decomposition of what we called
     "butterfly compounds" which are various
     salts of $Co(MoS_4)_2^{-2}$ and $Ni(MoS_4)_2^{-2}$ (7,8).
     In these compounds the promoter metal(s)
     (e.g. Co) and Mo are covalently bound.

II. "Decoration" of MoS_2 edge via a chemical
     reaction between MoS_2 with low-valent
     organometallic complexes such as $Co_2(CO)_8$
     (9,10). This decoration reaction, leading

to a rapid evolution of CO, is quite facile
at room temperature.

III. Thermal decomposition of metal amine
      thiomolybdates (MAT) in which the promoter
      metals and Mo, although not connected by
      any covalent bounding network, are molecu-
      larly associated with each other (11,12).

In this paper we restrict ourselves to the MAT-
derived catalysts. Specifically, we give an overview
of the performance of the bulk (unsupported) FeMo
sulfide prepared from MAT. This low-surface-area
catalyst shows a high HDN-to-HDS volumetric activity
ratio and is also active for HDA. While most of the
results are taken from our previous publications, some
new results are reported here.

EXPERIMENTAL

CATALYSTS

The precursor to the bulk FeMo catalyst used here
is bis (diethylenetriamine) iron thiomolybdate,
$Fe(H_2NCH_2CH_2NHCH_2CH_2NH_2)_2MoS_4$. The preparation of
this compound has been detailed elsewhere (11,12).
The preparation was typically carried out as follows:

First, 50 gm of $(NH_4)_2MoS_4$ were dissolved into 82
ml of diethylenetriamine (dien) in a one liter flask.
The resulting dark red solution was cooled to 0°C in
an ice bath and kept in the bath for the duration of
the experiment. In a separate flask, 42.84 gm of
$FeCl_2 \cdot 4H_2O$ were dissolved into 250 ml of distilled H_2O

and at least 25 ml of diethylenetriamine was added slowly to this Fe^{2+} solution to form $Fe(dien)_2^{2+}$. The resulting brownish $Fe(dien)_2^{2+}$ solution was allowed to cool at room temperature. The $Fe(dien)_2^{2+}$ solution was then added slowly, as aliquots, to the $(NH_4)_2MoS_4/$ dien solution with agitation for about 2 min after each addition. An orange red precipitate formed immediately. Distilled H_2O was added to increase the volume of the reaction mixture. The mixture was kept in the ice bath for at least 15 min until the reaction was completed. The precipitate was separated out by vacuum filtration through a Buchner funnel. The product was further washed with ethanol and dried under vacuum for 16-24 h. 93 gm of $Fe(dien)_2MoS_4$ were recovered.

A heterometallic metal sulfur complex of this kind may be called self-promoted, since the primary and promoter metals are molecularly associated with each other in a single complex. Unlike commercial catalysts, here the catalyst precursor is already in a sulfide form.

It is a simple matter to extend the above synthesis to prepare doubly-promoted catalysts; e.g., precursors of the form $M_xM'_{1-x}(amine)_{6/n}Mo_yW_{1-y}S_4$, where M and M' are different bivalent metal promoters, $0 \leq x$, $y \leq 1$, and n is the total number of nitrogen atoms in the amine chelating ligand.

Prior to use, the precursor compound needs to be sulfactivated; that is, it is thermally decomposed to remove its organic constituents in the presence of a sulfur-bearing stream. This was done as follows: The pelletized catalyst precursor was placed in a downflow fixed-bed reactor at 100°C at atmospheric pressure where it was purged for 1 h under N_2. Next a gas mixture containing 10% H_2S in H_2 was introduced into the reactor at a rate of 0.75 SCF/h for each 10 ml of catalyst precursor in the reactor. The reactor temperature was then raised to 325°C and kept at this temperature for 3 h. Following this the reactor was cooled to 100°C, the H_2S/H_2 gas flow was stopped and the reactor was purged with N_2 until room temperature was reached. The catalysts were then pressed into pellets and sized to 20-40 mesh granules.

Four commercial Al_2O_3-supported catalysts were tested to provide the base cases. Their properties are listed in Table 1. It should be pointed out that the support of catalyst D contains 1 wt % of SiO_2. These catalysts were presulfided at 360°C and ambient pressure for 1 h with a 10% H_2S-in-H_2 gas mixture.

TABLE 1 *Properties of commercial catalysts*

Catalyst	wt%			BET Surface Area, m^2/g	Pore Vol. cc/g
	MoO_3	CoO	NiO		
A	12.5	3.5	-	285	0.52
B	18.0	-	3.5	180	0.5
C	20.0	-	5.0	160	0.44
D	16.0	4.5	-	270	0.53

FEEDSTOCKS

The feedstocks used were two different light catalytic cycle oils (LCCO A and B) and a virgin gas oil.
Initial tests were done with LCCO A which contains 1.5
wt % total sulfur and 370 wppm total nitrogen. Subsequent tests were run on LCCO B containing 1.25 wt %
sulfur and 298 wppm nitrogen. The virgin gas oil,
with boiling range between 385 and 570°C, has 3 wt %
sulfur and 1000 wppm nitrogen. Details of the
feedstock properties can be found in (11-13).

ANALYTICAL

After purging with N_2, the liquid products were
analyzed for sulfur by X-ray fluorescence using a
Princeton Gamma-Tech Model 100 with a ^{55}Fe radioactive
source. Total nitrogen was analyzed by the Antek
combustion method, which utilizes chemiluminescent
detection of nitric oxide. Total carbon and hydrogen
were determined by Carbon^{13} and Hydrogen NMR.

Clay-gel separation (ASTM D2007) was used to
separate the liquid product into polars, aromatics,
and saturates. Following this, the saturate and
aromatic fractions were characterized by mass
spectrometry (ASTM D2786-71 for saturates, ASTM
D3239-76 for aromatics) according to the ring numbers
(Paraffins are classified as zero-ring naphthenes).
These procedures provided concentrations of seven
naphthenic ring-number fractions and 21 aromatic ring-

number fractions. The latter also included three
thiopheno-aromatic fractions (benzothiophenes,
dibenzothiophenes, and benzonaphthothiophenes).

It should be mentioned that the saturate fraction
contained a small amount of monoaromatics which could
not be analyzed by mass spectrometry. It was assumed
that the relative distributions of alkylbenzenes and
naphthenobenzenes in the saturate fraction are the
same as those in the aromatic fraction.

CATALYST EVALUATION

After presulfiding, the catalyst (typically 20-30
ml in 20/40 mesh granular form) was loaded into a
fixed-bed unit consisting of several independent
reactors disposed symmetrically in a common sand bath.
Each reactor was equipped with a calibrated feed
burette, a pump, a gas-liquid separator, and a product
liquid collector. The reactor was packed with catalyst in the central zone and 1/16" alundum in the fore
and aft zones. The FeMo catalyst has a packing density
of about 1.1 g/ml, vs. 0.8 g/ml for the commercial
catalysts.

The *volumetric* HDS and HDN activities were
determined by the rate constants of pseudo-second
order and pseudo-first order kinetics, respectively.
The selectivity of the catalyst for HDN relative to
HDS, S_N, is defined as the ratio of the rate constant
of HDN to that of HDS; that is, $S_N = K_N/K_S$.

Hydrogen consumption calculations were based on the increased hydrogen content in the liquid product and the hydrogen content in the gaseous products. The H_2S and NH_3 contents in the off-gas were calculated by the extents of HDS and HDN. It should be noted that the amount of light ends was very small; the liquid yields for all the runs were around 100% by weight.

CATALYST CHARACTERIZATION

The BET surface area of the FeMo catalyst after sulfiding is quite low, about 2 m^2/g. The pore-size distribution is bimodal, with peaks at 30 and 6000Å. The pore volume is 0.41 cc/g, which is predominantly (> 97%) contributed by the macropores (> 100Å).

Ho et al.(12) have characterized both the fresh and spent bulk FeMo catalysts by X-ray diffraction, transmission electron microscope, and Mössbauer. The key finding is that freshly sulfided bulk FeMo catalyst appears to be composed of an iron analog of the CoMoS material found by Topsøe et al. (14). This phase on oil partially transforms into pyrrhotite ($Fe_{1-x}S$) mixed with an MoS_2-like phase.

RESULTS

HDN/HDS OF LCCO

The activities and selectivities of the bulk FeMo sulfide and commercial catalysts for HDN and HDS are summarized in Tables 2 and 3. The results shown in

Table 2 were obtained with LCCO A, while those in
Table 4 were obtained with LCCO B. Evidently, the
FeMo catalyst gave a much higher S_N than catalysts A,
B, and C. This is mainly the result of the low HDS
activity of the FeMo catalyst.

TABLE 2   Catalyst Activity and Selectivity with LCCO A
            (325°C, 450 psig, 3000 SCF/B H_2, 1.5-6.0 LHSV)

Catalyst	K_N	K_S	S_N
A	0.55	10.7	0.05
B	1.26	9.9	0.13
FeMo	1.18	1.7	0.69

K_N: cc oil/cc cat.·hr.; K_S: cc oil/cc cat.·hr.·wt%

TABLE 3   Catalyst Activity and Selectivity with LCCO B
            (450 psig, 3000 SCF/B H_2, 2-5.5 LHSV)

	K_N		K_S		S_N	
	FeMo	Cat C	FeMo	Cat C	FeMo	Cat C
325°C	2	1.9	5	17	0.41	0.11
355°C	4.5	3.9	16	51	0.29	0.08

As is clear from Table 3, the selectivity advantage of
the FeMo catalyst shows little effect of temperature
over 325-355°C. The HDN and HDS apparent activation
energies with the FeMo catalyst are 20 kcal/mole and
28.8 kcal/mole, respectively, compared to 17.8 (HDN)
and 27.2 (HDS) kcal/mole for catalyst D. With either
catalyst, S_N decreases with increasing temperature,

presumably due to competitive adsorption and/or the higher HDS activation energy. Using the selectivity of the FeMo catalyst at 325°C as a tie point, one sees that catalyst C is more HDN selective than catalyst B.

HDA OF LCCO

The high HDN selectivity of the FeMo catalyst suggests that it may also be active for HDA. Indeed, as Fig.1 shows, FeMo was found to be more active than catalyst C at 325°C, 750 psig, and 1500 SCF/B.

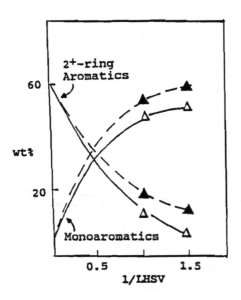

FIG. 1 *Concentrations (wt %) of 2+-ring and 1-ring aromatics vs. 1/LHSV for FeMo sulfide and catalyst C.* "———" for FeMo (Δ); "---" for catalyst C (▲).

CATALYST STABILITY

Accelerated aging following a pseudo-isoconversional policy was used to obtain information on the thermal stability of the FeMo catalyst. The results indicate that the catalyst is thermally stable (12).

HDN/HDS OF GAS OIL

Here the FeMo catalyst was tested against catalyst D. Figure 2 show the HDN and HDS kinetic plots
for these catalysts at 370°C, 1000 psig and 4000
SCF/B.

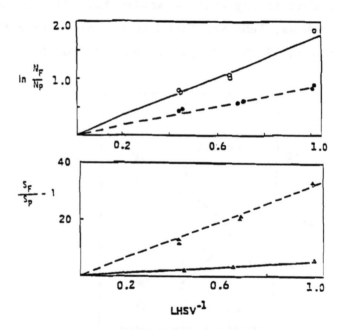

FIG. 2 *(a) First-order kinetic plots for HDN: "———"
for FeMo; "---" for catalyst D. N_F and Np are wppm of
nitrogen in the feed and liquid product, respectively.
(b) Second-order kinetic plots for HDS. S_F and Sp are
wt % of sulfur in the feed and product, respectively.*

The regression results are summarized in Table 4.
Within the range of conditions employed, the FeMo
catalyst is about twelve times more selective for HDN
than catalyst D (CoMo). This is consistent with the
results shown in Table 2 (FeMo vs. catalyst A).

TABLE 4   *HDN/HDS Activity and Selectivity*

	K_S	K_N	S_N
Catalyst C	10.56	0.85	0.08
FeMo	1.72	1.71	0.99

Figures 3 and 4 show the temperature dependencies of K_N and K_S for both catalysts at 1.0 LHSV and 7.0 MPa, respectively. The apparent activation energies for HDN and HDS on the FeMo catalyst are, respectively, 30 and 65 kcal/mole, while those on catalyst D are 23 and 55 kcal/mole, respectively.

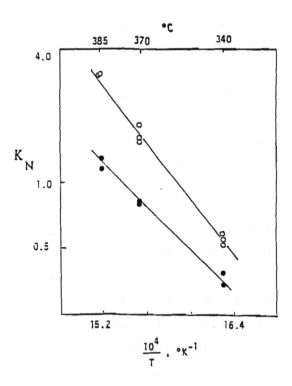

FIG. 3 *HDN Arrhenius plots for bulk FeMo sulfide (o) and catalyst D (●).*

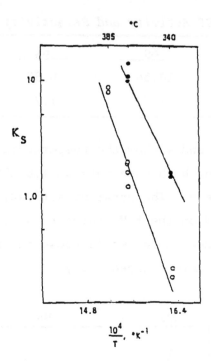

FIG. 4 *HDS Arrhenius plots for bulk FeMo sulfide (o)*
and catalyst D (•).

The pressure dependence can be correlated by
power law kinetics of the form

$$K_N = k_n P^{\alpha}_{H_2} \qquad K_S = k_s P^{\beta}_{H_2}$$

Over the conditions used, the values of α and β are
slightly different for different catalysts, but $\alpha =$
1.0 and $\beta = 0.3$ provide acceptable representations in
all cases. That $\beta < \alpha$ is not surprising (3). As shown
in (13), the selectivity advantage of the FeMo catalyst was seen to hold over a wide range of conditions.
In particular, the FeMo bulk sulfide becomes more HDN
selective than catalyst D at high pressures or at high
temperatures.

HDA OF GAS OIL

A. LHSV Effects on Total Homocyclic Aromatics

Figure 5 plots the concentrations (wt %) of total aromatics and saturates against 1/LHSV for FeMo and catalyst D at 370°C and 1000 psig. The curves in this figure and those in all figures to follow are arbitrarily drawn.

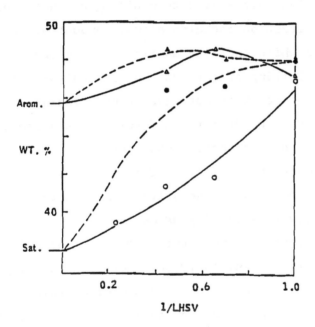

FIG. 5 *Concentrations (wt %) of total homocyclic aromatics and saturates vs. 1/LHSV. "----" for catalyst D (●, ▲); "———" for FeMo (o, Δ)*

With either catalyst, one immediately sees that a contact time (1/LHSV) of one hour is too short to achieve a *net* decrease in homocyclic aromatics. But at this condition, CoMo achieved a 97% HDS and a 57%

HDN while FeMo gave a 83% HDS and a 83% HDN. Apparently, the increases in homocyclic aromatics and saturates are primarily due to heteroatom removal. In other words, heteroatom removal is the primary reaction occurring on either catalyst. The higher saturate buildup with catalyst D results from the catalyst's high HDS activity and the fact that the feed contains far more sulfur than nitrogen. Note that the saturate fraction contains a small amount of monoaromatics which increase as reaction proceeds.

The buildup of saturates over catalyst D is rather fast at first but then becomes increasingly slow as HDS gets deeper and deeper. The accompanying change in total aromatics goes through a maximum and then appears to level off. This suggests that after removing the bulk of the sulfur, the catalyst was rather slow in attacking aromatics.

In contrast, the saturate make on the FeMo catalyst is low initially but increases steadily with time. The concentration of homocyclic aromatics goes through a maximum and then falls. The total aromatics with FeMo at 1/LHSV = 1.0 h was lower than that with catalyst D. Apparently, after removing a substantial amount of nitrogen, the FeMo catalyst becomes effective in hydrogenating aromatics, despite the fact that there are still an appreciable amount of organosulfur compounds.

The behavioral differences between the two catalysts can be seen more clearly by examining the distributions of individual aromatic ring-number lumps.

B. LHSV Effects on Individual Aromatic Lumps

Referring to Fig.6, the reduction of multiring aromatics proceeds in a stepwise manner and virtually stops at the monoaromatic step, resulting in a buildup of monoaromatics. In order to achieve an appreciable net reduction of monoaromatics, the space time will have to be much longer than one hour. As will be seen later, over the conditions tested, the hydrogenation reactions are not equilibrium controlled.

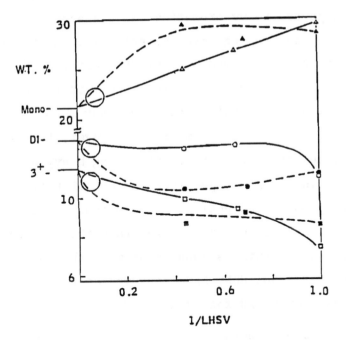

FIG. 6 Concentrations (wt %) of monoaromatics, diaromatics, and 3+-ring aromatics as functions of 1/LHSV. "----" for catalyst D (solid symbols); "——" for FeMo (open symbols).

The buildup of monoaromatics was faster and larger with catalyst D, and there is a rapid initial reduction of 2- and 3^+-ring aromatics in the absence of significant HDN. After this initial period, further reduction of these heavy aromatics became quite slow (actually the amount of diaromatics increased slightly). Apparently, catalyst D had difficulties hydrogenating heavy refractory aromatics, although it was able to hydrogenate the reactive ones. That the catalyst is supported on SiO_2-Al_2O_3 may be a factor here: the support offers acidic sites on which these large and heavy aromatics can preferentially adsorb (15) and undergo hydrogenation most probably via protonation followed by hydride transfer (16).

The bulk FeMo catalyst exhibited just the opposite behavior. Initially, there was not much reduction in 2- and 3^+-ring aromatics. However, the catalyst became more effective in attacking heavy aromatics *after* removing a significant amount of nitrogen, presumably due to competitive adsorption. This attack occurred at HDS levels much lower than those attained by catalyst D. This is consistent with the observation that at a 50% HDN, FeMo consumed about 300 SCF/B of H_2, vs. 600 SCF/B for catalyst D.

C. Temperature Effects

Figure 7 shows the effects of temperature on the reduction of total homocyclic aromatics at 1000 psig and 1.0 LHSV.

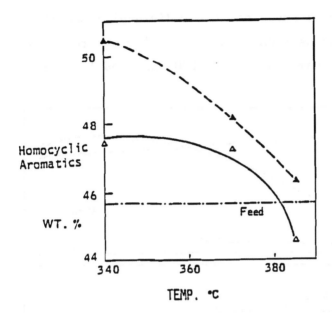

FIG. 7 *Temperature effects on the reduction of total homocyclic aromatics. "---" for catalyst D (▲); "——" for FeMo (△).*

The data for the FeMo catalyst suggest that the maximum increase in homocyclic aromatics might have occurred at around 340°C. With catalyst D, the maximum most likely have occurred at a temperature below 340°C. Neither catalyst was equilibrium limited. At 385°C FeMo could reduce the total aromatics to a level below that in the feed while catalyst D could not. At this temperature both catalysts achieved comparable HDS levels (100% for catalyst D and 96% for FeMo). Thus the difference in HDA activity between the two catalysts should be attributed to the higher HDN activity of the FeMo catalyst (96% HDN with FeMo and 71% HDN with catalyst D).

Figure 8 shows how the individual homocyclic aromatic lumps (mono-, di-, and 3⁺-ring aromatics) respond to temperature at 1000 psig and 1.0 LHSV.

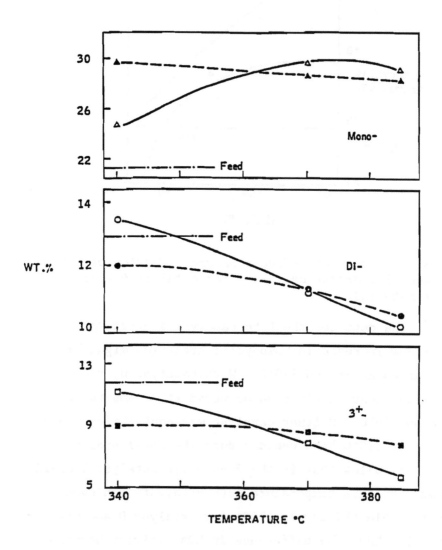

FIG. 8 *Effects of temperature on the reduction of individual aromatic lumps. "----" for catalyst D (solid symbols); "———" for FeMo (open symbols).*

Over catalyst D, increasing temperature leads to rather slow and more or less parallel reductions of all three aromatic lumps. At 340°C both the concentrations of di- and 3+-ring aromatics are lower than those observed with the FeMo catalyst, indicating that at low temperatures catalyst D is more effective than FeMo in hydrogenating some reactive (e.g. sterically unhindered) multiring aromatics. The FeMo catalyst, on the other hand, shows a stronger temperature dependence and hence is more effective for hydrogenating multiring aromatics at high temperatures.

In Fig.9 the total concentration of 3+-ring aromatics plus polars is plotted against temperature at 1000 psig and 1.0 LHSV. The numbers indicated on the figure are nitrogen contents (wppm) of the product liquids at 385°C. As seen, once the bulk of nitrogen is removed, the FeMo catalyst becomes more effective than catalyst D for attacking the potent coke precursors. This behavior is consistent with that shown in Fig. 6.

It should be mentioned that with either catalyst a reduction in monoaromatics cannot be achieved at 385°C. A related question is whether there is inter-conversion between different types of monoaromatics. As Fig. 10 shows, the ratio of alkylbenzenes to naphthenoaromatics at 1000 psig and 1.0 LHSV remained virtually constant over 340-385°C, suggesting that ring opening of naphthenomonoaromatics to alkyl-benzenes did not appear to occur.

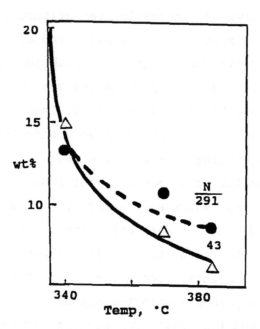

FIG. 9 *Effects of temperature on the reduction of 3+-ring aromatics plus polars.* *"----" for catalyst D (●); "——" for FeMo (o).*

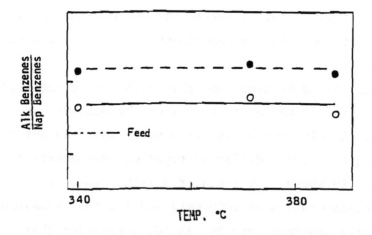

FIG. 10 *Effect of temperature on the ratio of alkyl-benzenes to naphthenobenzenes.* *"----" for catalyst D (●); "——" for FeMo (o).*

D. *Pressure Effects*

The H_2 pressure dependence of overall HDA is shown in Fig. 11. The data were obtained at 340°C and 1.0 LHSV. As expected, the concentrations of total homocyclic aromatics decreased with increasing pressures. Catalyst D, being preoccupied with HDS, showed a weaker pressure dependence.

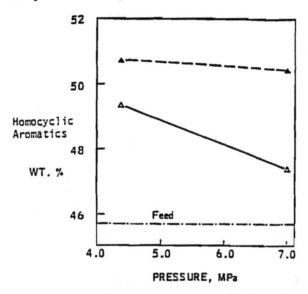

FIG. 11 *Effect of H_2 pressure on hydrogenation of total homocyclic aromatics."----" for catalyst D (▲); "——" for FeMo (△).*

DISCUSSION

The bulk FeMo catalyst has been evaluated against four different commercial catalysts with three different feedstocks over a broad range of conditions. The results all point to the high volumetric selectivity of this low-surface-area catalyst toward HDN. Here we put together simple, lumped reaction networks to

contrast the bulk FeMo sulfide with a typical commercial catalyst. The reaction network can be schematically represented by a network of interconnected
tanks, as depicted in Fig.12. In each tank, the dotted
line represents the amount of a particular lump (e.g,
total sulfur, 2^+-ring aromatics, etc.) in the feed,
whereas the solid line represents the amount of that
lump at a particular set of reaction conditions.

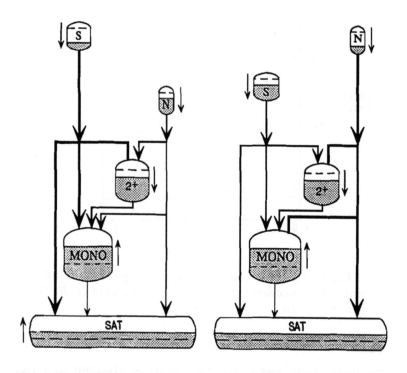

FIG. 12 *Schematic diagrams showing the lumped
hydroprocessing network: (a, left) a typical
commercial catalyst; (b, right) bulk FeMo sulfide.*

As can be seen, while heteroatom removal is the
primary reaction on both catalyst, the commercial
catalyst has HDS as its principal pathway, Fig.12a.

In the case of FeMo, HDS becomes less important, Fig. 12b. Here we assume that the conditions are such that the HDA reactions are irreversible. The bulk FeMo sulfide is a good HDA catalyst due to its high HDN selectivity. However, the downside to this property is that the active sites on FeMo have such a strong affinity for nitrogen compounds that significant HDA does not occur until the bulk of nitrogen compounds is removed. On the other hand, active sites on the commercial catalyst appear to have less of an ability to discriminate the nitrogen and aromatic compounds. As a result, HDA is less vulnerable to nitrogen inhibition. The acidity of the oxide support also plays a role here: it can facilitate HDA of some large and presumably sterically unhindered aromatic compounds.

Finally, some conjectures regarding the characteristics of the active sites on the FeMo sulfide catalyst seem to be in order. That the activation energy for HDN on the FeMo catalyst is higher than that on commercial catalysts suggests that the FeMo catalyst probably provides more sites where nitrogen heterocycles can be preferentially adsorbed and/or hydrogenated. The low HDN selectivity of the commercial catalysts may be partially attributed to the presence of the Al_2O_3 support.

It has been well recognized that HDN involves two consecutive steps: hydrogenation followed by hydro-

genolysis (3).   In HDN of basic nitrogen compounds
(e.g. quinolines) over conventional catalysts, the
weight of evidence indicates that hydrogenolysis
appears to be the rate-limiting step under most
commercial conditions. As for multiring nonbasic
nitrogen compounds (e.g. carbazoles), indications are
that the rate of hydrogenation can be much slower than
that of hydrogenolysis (3). One may further *speculate*
that the FeMo and commercial catalysts probably have
more or less the same ability for the adsorption and
hydrogenation of basic nitrogen heterocycles. What
makes the FeMo catalyst more selective for HDN might
be the catalyst's stronger affinity for nonbasic
nitrogen heterocycles. If so, those sites are not
highly acidic in nature. This is consistent with the
observation that the FeMo catalyst is quite active
even though it is exposed to an environment having
high NH_3 and low H_2S. Further work addressing this
fundamental issues is needed.

CONCLUDING REMARKS

Summarizing, the basic idea behind our approach
is that the formation of catalytically active phases
can be facilitated by preparing molecular catalyst
precursors in which Mo, promoter metal(s), and sulfur
are intimately mixed at the molecular level. This
approach seems to offer a greater control of catalyst
activity and selectivity than conventional catalyst
preparation chemistry. Further results in this respect

are forthcoming (17).

ACKNOWLEDGMENT

Thanks are due to R. L. Seiver for providing Figure 1.

REFERENCES

1. Gates, B.G.; Katzer, J.R.; and Schuit, G.C.A.;
    "Chemistry of Catalytic Processes" McGraw-Hill,
    New York, 1979.

2. Satterfield, C.N.; "Heterogeneous Catalysis in
    Industrial Practice," 2nd Ed. McGraw-Hill, New
    York, 1991.

3. Ho, T.C.; Catal. Rev. - Sci. Eng. 30, 117, 1988.

4. Prins, R.; DeBeer, V.H.J.; Somorjai, G. A.;
    Catal. Rev. - Sci. Eng., 31, 1, 1989.

5. Chianelli, R.R.; Ruppert, A.F.; Behal, S.K.;
    Kear, B.H.; Wold, A.; and Kershaw, R.; J. Catal.,
    92, 56, 1985.

6. Stiefel, E.I.; Halbert, T.R.; Coyle, C.L., Pan,
    W.-H; Ho, T.C.; Chianelli, R.R.; and Daage, M.;
    Polyhedron, 8, 1625, 1989.

7. Stiefel, E.I.; Pan, W.-H.; Chianelli, R.R.; and
    Ho, T.C.; U.S. Patent 4581125, 1986; to Exxon
    Research and Engineering Company.

8. Pan, W.-H.; McKenna, S.T.; Chianelli, R.R.;
    Halbert, T.R.; Hutchings, H.H.; and Stiefel,
    E.I.; Inorg. Chim. Acta 97, L17, 1985.

9. Halbert, T.R.; Stiefel, E.I.; Chianelli, R.R.;
    and Ho, T.C.; U.S. Patent 4839326, 1989; to Exxon
    Research and Engineering Company.

10. Halbert, T.R.; Ho, T.C.; Stiefel, E.I.;
    Chianelli, R.R.; and Daage, M.; J. Catal., 130,
    116, 1991.

11. Ho, T.C.; Young, A.R.; Jacobson, A.J.; and
    Chianelli, R.R.; US Patent 4591429, 1986; to
    Exxon Research and Engineering Company.

12. Ho, T.C.; Jacobson, A.J.; Chianelli, R.R.; and
    Lund, C.R.F.; J. Catal., 138, 351, 1992.

13. Ho, T.C.; Ind. Eng. Research, 32, 1568, 1993.

14. Topsøe, H.; Clausen, B.S.; Wivel, C.; and Mørup,
    S.; J. Catal., 68, 433, 1981.

15. Ho., T.C.; Katritzky, A.R.; and Cato, S.J.; Ind.
    Eng. Chem. Res., 31, 1589, 1992.

16. Pines, H.; "The Chemistry of Catalytic Hydrocar-
    bon Conversions, Academic Press," pp 99-100,
    1981.

17. Ho, T.C.; Chianelli, R.R.; and Jacobson, A.J.; to
    be published.

21 Fluidized Catalytic Cracking of Hydrotreated Charge Stock for Naphtha Sulfur Reduction

James Mudra

Texaco Research & Development
Port Arthur, Texas

ABSTRACT

New environmental laws will require the gasoline composition to change dramatically in order to reduce automotive emissions. Auto/Oil studies have indicated that reducing sulfur will lower the emissions of hydrocarbons, NOx, and CO. Since a large portion of the sulfur in the gasoline pool comes from the FCC naphtha, reducing the FCC naphtha sulfur level will lower the sulfur level in the gasoline pool. Hydrotreating the feedstock to the FCC is an effective means of lowering FCC naphtha sulfur. Pilot unit experiments using hydrotreated feeds from Arabian and Alaskan North Slope crudes have shown that the sulfur level in the DB naphtha is reduced dramatically depending on the hydrotreating severity. Cutting the heavy naphtha endpoint will also help lower the sulfur level.

INTRODUCTION

With the passage of the Clean Air Act (CAA) by Congress in 1990, gasoline composition will undergo a drastic change to reduce automobile emissions in order to meet future environmental requirements. The CAA requires that the oxygenated fuels be used in certain areas, and that the emissions of hydrocarbons, NOx, and CO must be reduced. Other limitations in the CAA include one percent or less benzene and no more than 20% aromatics in the gasoline pool. The results of Auto/Oil studies have indicated that reducing any properties in the gasoline pool would only reduce, at most, two of the three emissions. The one exception is sulfur. Since sulfur can affect the catalyst activity in the catalytic convertor of an automobile, a reduction in sulfur can reduce the emissions of hydrocarbons, NOx, and CO. The prevailing opinion in the refining industry is that a reduction in sulfur content in gasoline will be required.

In California, reducing the sulfur content in the gasoline pool will be required according to the California Air Resource Board (CARB). The Phase 2 regulations, as required by CARB, will allow a maximum sulfur limit of 40 ppm in the gasoline pool if the averaging method is not used. Using the averaging method, the sulfur gasoline content limit is even lower at 30 ppm. With many other states considering adopting California's CARB Phase 2 regulations, sulfur reduction will be mandatory in many parts of the United States.

Table I and Table II summarize the U.S. gasoline blending components and fluidized catalytic cracking (FCC) feedstock utilization[1,2]. The average sulfur content in today's gasoline is 143 ppm for premium grade and 384 ppm for regular grade. Comparing these numbers with CARB's required sulfur amount, a very large reduction in sulfur needs to be achieved. The main contributor of sulfur to the gasoline pool is the naphtha products from the FCC units. Since 14 - 47 vol% of the gasoline pool is made up directly from the FCC unit and another 7 - 23 vol% is contributed indirectly, mostly from alkylation, the products from the FCC cannot be easily rejected in order to lower the sulfur content of the gasoline pool. Currently 68% of the FCC feedstock in the U.S. is not hydrotreated and must be treated by some method to lower the sulfur level.

Options are currently being explored to determine ways to lower the sulfur level from the FCC products. These include: new catalyst development, undercutting the heavy naphtha gasoline, and treating the feed or product of the FCC. Other options, such as processing more paraffinic or naphthenic crudes, are unrealistic since most crude slates are becoming more aromatic. New catalyst development is expected to help lower the sulfur level, but today's catalyst will not meet the future sulfur levels in the gasoline pool[3]. Undercutting the heavy end of the naphtha gasoline will help lower the sulfur and aromatics content, but again this route does not solve all the regulatory concerns completely. Some data will be presented where the ninety percent point (T90) cut for naphtha gasoline is lowered to show the effects of sulfur reduction. Finally, treating the feed or product from an FCC is a viable option that can be done. However, Gialla et al. stated that desulfurization of the FCC gasoline by hydrofinishing units is only a short term solution[4].

Table I: *U.S. Gasoline Blending Components*

Component (vol%)	Premium Grade	Mid-grade	Regular Grade
FCC Naphtha	14.1	45.6	46.7
Reformate	42.1	22.2	21.8
SR Naphtha	0.5	1.6	5.0
Isomerate	2.2	3.3	4.2
Coker Naphtha	0.214.1	0.4	0.8
Hydrocracked	0.21.1	2.3	2.6
Alkylate	23.0	11.3	7.4
MTBE	2.7	1.2	0.2
n-Butane	3.7	3.5	2.9
Others	10.4	8.6	8.4
% of Unleaded Pool	27.5	8.1	64.4
Sulfur (ppm)	143	375	384

1989 Data - NPRA Survey

Hydrotreating the feed to an FCC will result in benefits that are well balanced in complying with the new environmental legislation. Naphtha yields will increase at the expense of dry gas and coke. The sulfur content in the naphtha gasoline, gas oils, and flue gas will all decrease. Nitrogen will also

Table II: *U.S. FCC Utilization*

	Utilization (bpd)	Sulfur (wt%)
Unhydrotreated	3,072,700	1.02
Hydrotreated	1,454,115	0.37
Total	4,526,815	

1989 Data - NPRA Survey

decrease which will help in the catalyst selectivity and activity. The aromatics content will also decrease which will help comply with the CAA. As stated earlier, 68% of the feed to the FCC is unhydrotreated while the remainder is hydrotreated to an average sulfur level of 0.37 wt%. The portion that is not hydrotreated has an average sulfur level of 1.02 wt%. The sulfur level in this portion will have to be reduced. The major drawback of hydrotreating the FCC feedstock is the intensive capital investment for the refiner.

EXPERIMENTAL BACKGROUND

The sulfur data was derived from a series of seven hydrotreating experiments which were done at the Texaco Research & Development - Port Arthur facility. Four of these experiments were specifically designed to determine the desulfurization effects that severe hydrotreating has and the possibility of producing low sulfur naphtha. These four experiments represent the basis of this report, and most quantitative comparison will be done using the four experiments. The other experiments were performed prior to the passage of the CAA but were designed to determine the affects that mild hydrotreating has on the FCC's product yield and quality.

Equipment

Most of the experiments were performed on the Miniature Riser Unit (MRU). This FCC pilot unit can typically charge an average of 1 l/hr of feedstock. The feed oil is injected at the base of a riser where it immediately contacts the regenerated catalyst stream. The catalyst and oil mixed with lift nitrogen travel up the riser as the feed is cracked to products. The catalyst and oil vapor mixture enters the reactor where the vapor products pass overhead and the catalyst falls to the bottom of the reactor and then, by gravity, to the stripper. Entrained hydrocarbons are removed from the catalyst in the stripper by a flow of nitrogen. The catalyst is lifted by nitrogen to the regenerator where air is used to burn the carbon off the catalyst at an elevated temperature, approximately 1330 F. After the carbon is removed from the catalyst, the catalyst then flows down the standpipe and mixes with the feed oil. Because the MRU is not a heat balanced unit, the riser temperature, regenerator temperature and catalyst circulation can be adjusted independently.

The reactor effluent is cooled by a separator system. Any product gas not condensed in the separator is metered by a gas meter and a slip stream is analyzed by an on-line gas chromatograph for hydrocarbons, hydrogen, hydrogen sulfide and nitrogen. The hydrocarbon liquid product from the separator is sent to a total liquid product graduate. A sample of the total liquid product is analyzed by simulated true boiling point (TBP) distillation in a gas chromatograph. The remaining total liquid product is sent to a high efficiency batch fractionation columns where the depentanized light naphtha (115-250 F TBP), heavy naphtha (250 - 430°F TBP), light cycle gas oil (430-670°F TBP) and heavy cycle gas oil (670°F+) fractions are obtained for product testing.

Regenerator flue gas samples are analyzed by an on-line gas chromatograph for nitrogen, oxygen, carbon monoxide and carbon dioxide. Regenerator flue gas is also monitored for concentrations of SO_2 and SO_3 by a two-stage absorption system or by a SO_2 analyzer.

The other pilot unit that was used was a 5 - BPD FCC pilot unit. The operation of this unit is similar to the MRU with the main exception that the 5 - BPD unit is heat balanced. This unit was only used in one of the mild hydrotreated experiments.

Charge Stock

Sulfur distribution in the FCC products is not only dependent on the initial feed sulfur but also the refinery feed source. If the feed is hydrotreated, the operating conditions of the hydrotreater will affect the FCC product sulfur distribution as well as the conversion and yields. It is not our intention to determine the optimum hydrotreating method or operating conditions, but to determine the effect of hydrotreating on the FCC. Data within this report utilized a fixed bed hydrotreater and an ebullated bed hydrotreater (Texaco's T-STAR^{SM} process).

In the experiments that were designed for low sulfur naphtha, the charge stocks that were used were a blend of Alaskan North Slope (ANS) heavy vacuum gas oil (HVGO) and heavy coker gas oil (HKGO), Arabian HVGO and H-Oil$^{®}$, Arabian HVGO and ANS HKGO, and 100% Arabian HVGO. The gravity, sulfur, and nitrogen properties of these unhydrotreated feeds are shown in Table III. The H-Oil$^{®}$ process (jointly licensed by Texaco and HRI) is a hydrocracker that is used to upgrade vacuum residuum fractions. Feed properties for the hydrotreated experiments are shown in Table IV. Hydrotreating the feed reduces the overall sulfur level significantly. The sulfur level for the ANS experiment dropped from 1.45 wt % to the 560 - 2311 ppm level, a decrease of 84 - 96 %.

Table III: *Unhydrotreated Feed Properties*

	ANS: 87.5 vol% HVGO 12.5 vol% HKGO	Arabian: 80 vol% HVGO 20 vol% H-Oil®	Arabian: 100 % HVGO
API Gravity	19.7	19.3	20.4
Sulfur (wt %)	1.45	2.55	3.04
Nitrogen (wt%)	1814	1163	994

Table IV: *Hydrotreated Feed Properties*

Initial Sulfur (wt%)	Hydrotreated Sulfur (wppm)	Gravity	Nitrogen (wppm)
1.45	560 - 2311	22.2 - 26.2	557 - 1201
2.55	714 - 6100	24.4 - 29.9	N/A
3.04	525 - 2000	29.4 - 30.5	N/A

HYDROTREATING EFFECTS

Hydrotreating the feed to the FCCs has been done since the mid-70's. In the past, the reason for hydrotreating has mainly been for upgrading heavy oil, but because of the CAA and the Phase II regulations, hydrotreating is becoming an environmental solution. Other benefits besides the environmental impact have been known and will be briefly discussed.

General Hydrotreating Benefits

Hydrotreating the feed caused a general improvement in feed quality. The conversion increased (Figure 1). A more severe hydrotreating caused a higher conversion (100 vol% - gas oil vol%) at the same riser outlet temperature. At lower riser temperature, the hydrotreating effect on conversion was greater, but at that point maximum conversion had not been reached. Both the coke yield and dry gas yield were reduced (Figure 2 and 3) as hydrotreating severity increased. The lower coke yield would allow a lower regenerator temperature which in turn could allow a refiner to utilize a higher catalyst to oil ratio. This would cause a more selective cracking in the riser to occur. The total DB naphtha yield increased, as expected (Figure 4) as hydrotreating severity increased. In this example the naphtha peak point occurred around 74 vol% for the unhydrotreated feedstock. Hydrotreating shifted this point to a greater conversion. The moderate hydrotreating curve had a naphtha peak point around 79 vol% conversion. Hydrotreating the FCC feedstock reduced the sulfur concentration in all streams from the FCC. The level of sulfur in all the product streams including H_2S, the naphthas and the gas oils was lower. Sulfur oxide (SOx) emissions were also reduced. Hydrotreating the FCC feedstock would help refiners comply with low sulfur gasoline mandates and SOx emission restrictions, while increasing the quantity of DB naphtha.

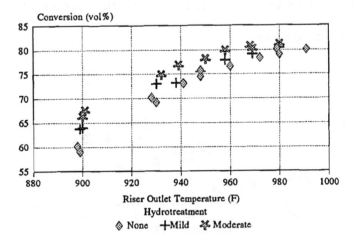

Figure 1: *Conversion vs. Riser Outlet Temperature*

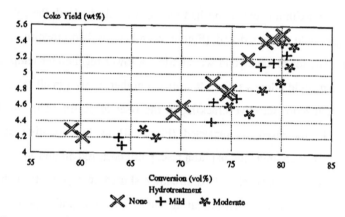

Figure 2: *Coke vs. Conversion*

Basics of Sulfur Distribution in the FCC Process

The sulfur in the feedstock contained numerous species of sulfur, many bonded to the long chain molecules in the VGO. The FCC process eliminated 40% or more of the sulfur in the liquid process. Overall, 40-50% of the initial feed sulfur was converted by the cracking reactions to H_2S. Around 5% of the

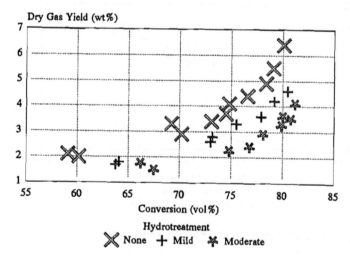

Figure 3: *Dry Gas vs. Conversion*

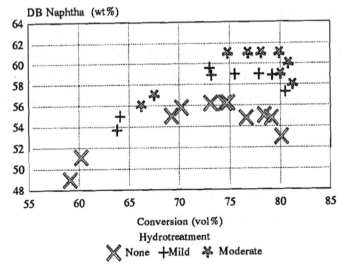

Figure 4: *DB Naphtha vs. Conversion*

sulfur became SOx emissions from the burning of the coke in the regenerator. The balance remained in the liquid product. The average gas oil product sample contained over 90% of the sulfur that occurs in the liquid product. The remaining 10% of the sulfur in the liquid product occurred in the naphtha and would have to be reduced in order to comply with the new regulations. These values could change with feedstock and FCC operating conditions.

The general groups of sulfur that existed in the FCC liquid products include: mercaptans, sulfides, thiophenes, benezothiophenes, and other poly-nuclear aromatic compounds containing sulfur. The level of desulfurization reactivity decreased from left to right. If sufficient reactivity exists, mercaptans and sulfides could be converted into H_2S. Since aromatics are not easily converted in the FCC, benzothiophenes and other aromatic containing sulfur components remain unchanged in the FCC process.

DB Naphtha Sulfur

The sulfur in the naphtha gasoline represented 10% or less of the feed sulfur. Plotting the initial feed sulfur versus the DB naphtha sulfur for unhydrotreated and hydrotreated feedstocks on a logarithmic scale yielded a straight line, indicating a non-linear function (Figure 5). The major benefit of hydrotreating the feed for sulfur reduction can be seen in this graph. The amount of sulfur was dependent on the operating conditions of the hydrotreater. The higher severity in the hydrotreater would result in a greater reduction in the overall sulfur level in the FCC feedstock. The benefits of hydrotreating for sulfur reduction in the FCC feed could already be realized even before the feed was charged to the FCC.

Plotting the hydrotreated ANS, Arabian, and Arabian/ANS blend DB naphtha sulfurs versus FCC feed sulfurs data (below the initial feed sulfur of 2500 ppm), yielded Figure 6. Here, individual trends dependent on the feed type were shown to have similar slopes. The Arabian HVGO/ANS HKGO blend had the highest amount of sulfur which could be attributed to the fact that

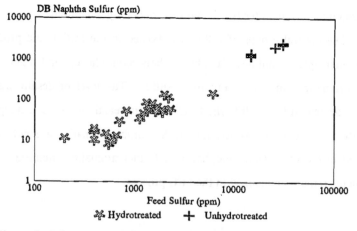

Figure 5: *DB Naphtha Sulfur vs. Feed Sulfur*

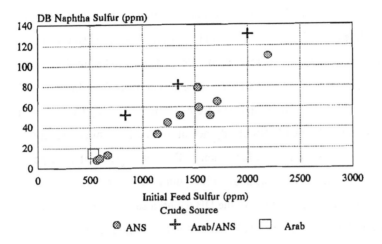

Figure 6: *DB Naphtha Sulfur vs. Feed Sulfur*

this experiment had one of the highest initial feed sulfur levels of 3.04 wt %.

DB naphtha sulfur for unhydrotreated feed usually was reduced by 90 - 92 % by going through the FCC. Combining the sulfur reduction due to the FCC and the hydrotreater, the Arabian/ANS blend sulfur level could be reduced by over 99 wt% in the DB naphtha stream. No experiments for the hydrotreated ANS, Arabian, and Arabian/ANS experiments were attempted above 2500 ppm feed sulfur because initial results had shown that compliance to the new regulations would require the initial feed sulfur to be below the 2500 ppm sulfur level. The average hydrotreated FCC feedstock in the U.S. is 0.37 wt% indicating that some of the current hydrotreaters would need to be operated at a more severe desulfurization condition or additional FCC pretreaters will need to be built unless blending with low sulfur components could compensate for the higher FCC sulfur products.

Most of the experimentation in our study was performed on various samples derived from the ANS HVGO/HKGO feedstock. Under the most severe feedstock hydrotreated case that was tested, the sulfur in the DB naphtha was below 10 ppm. The ANS experiment (which contained 14 runs) displayed a very linear relationship of DB naphtha sulfur versus feed sulfur until the feed sulfur level increased to 1500 ppm. Beyond the 1500 ppm feed sulfur level, the ANS experimental data showed some scatter away from linearity, while the highest sulfur feed varied upwards from the line. This upward trend can also be seen in the Arabian/ANS blend, indicating that there might be an increasing slope past a certain feed sulfur content. For our experiments, the point where the non-linearity began was above 1500 ppm feed sulfur. Below 1500 ppm feed sulfur, the feedstock sulfur and the DB naphtha sulfur had a linear relationship that was dependent on the crude source and the level of hydrotreating.

The point where the slope becomes non-linear would be important in determining the level of hydrotreating that would be needed to produce a naphtha product at a particular sulfur level. Because Figure 5 shows that the overall hydrotreated and unhydrotreated transition as a non-linear relationship, the linear relationship that was seen below 1500 ppm could not continue into the higher sulfur region. Unfortunately, additional experiments were not performed above the 2500 ppm sulfur level which would have been beneficial in determining the non-linear region. When estimating for a 50 ppm sulfur DB naphtha in the ANS experiments at the chosen FCC operating conditions, the hydrotreated feed sulfur would need to be about 1475 ppm. This is not an unreasonable sulfur level in the DB naphtha to comply with the Phase II regulations of 30 ppm in the gasoline pool. With reformulated gasoline, the new components such as MTBE would likely contain a very low level of sulfur, if any. This would allow the sulfur level in the FCC DB naphtha to be at a higher level and still meet the regulation mandates. A DB naphtha sulfur level of 100 ppm would require the initial feed sulfur to the FCC to be around 2400 ppm assuming a linear relationship for the ANS experiment. Using a non-linear slope, the initial feed sulfur to the FCC could be around 2100 ppm.

Reduction of feed sulfur to the FCC can be accomplished by incorporating existing units in refineries with a excess hydrotreating capacity. If a hydrocracker already exists in the refinery but could not be operated at severe enough conditions to meet the CAA or Phase II requirements, the hydrotreater could be used in series with another unit which could be operated at a moderate severity. Figure 7, shows the ANS and Arabian data at a constant FCC operating conditions, but different hydrotreating conditions were used to obtain the FCC feeds. Two other runs were done using a mixture of Arabian/H-Oil$^{®}$. The H-Oil$^{®}$ process is a hydrocraking process licensed by Texaco and HRI which allows for resid upgrading. One of the Arabian/H-Oil$^{®}$ mixtures to the pilot unit had a high feed sulfur level while the other was hydrotreated severely. Assuming a straight linear relationship, the resulting slope of the DB naphtha versus feed sulfur was significantly lower than the ANS - Arabian line. A 100 ppm sulfur level in the DB naphtha would require an initial feed sulfur near 4000 ppm. Without data between the two points, the

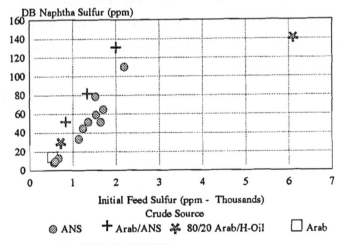

Figure 7: *DB Naphtha Sulfur vs. Feed Sulfur*

determination where the non-linear region begins could not be found. The main objective was to demonstrate the utilization of the hydrocracker in conjunction with a hydrotreater.

Light and Heavy Naphtha

Table V splits the total DB naphtha into light naphtha (115 - 250 F) and heavy naphtha (250 - 430 F) streams. The volume of the FCC light naphtha increased as hydrotreating severity was increased for the ANS and Arabian/H-Oil$^{®}$ and remained the same for the 100% Arabian HVGO. The heavy naphtha volume percent seems to be a function of feedstock. Examining the sulfur levels between the light and heavy naphthas, the light naphthas had a significantly lower level of sulfur regardless of hydrotreating severity. In the most severe case of hydrotreating, the ANS light and heavy naphtha sulfur was 5.1 ppm and 15.7 ppm, respectively. This was a drastic reduction from the unhydrotreated level of around 600 ppm and 2000 ppm for light and heavy naphtha, respectively. Another severe hydrotreating case was done for the 100% Arabian HVGO. The light naphtha had a sulfur level of 3.0ppm, and the heavy naphtha sulfur level - 430 cut was very high in both the unhydrotreated and hydrotreated naphthas. 26.0 ppm. This demonstrated that various feedstocks could be severely hydrotreated to meet the CARB sulfur requirements.

Undercutting the Heavy Naphtha

The CARB Phase 2 regulations require that the gasoline T90 point be at 290 F for the averaging limit. This may cause the T90 in the FCC naphtha stream to be lowered. Several of the runs had the heavy naphtha portion undercut from 430 to 400 F, as shown in Table VI. The sulfur concentration in the 400 - 430 cut was very high in both the unhydrotreated and hydrotreated naphthas. The 250 - 400 portion for the unhydrotreated runs had a sulfur level in the 2500 - 2800 ppm range while containing 80% or more of the heavy naphtha. The 400 - 430 portion of the unhydrotreated run had a very high sulfur level of over 10,000 ppm. This trend was also seen in the hydrotreated cases. The 400 - 430 fraction contained 35 - 45% of the sulfur that was found in the overall 250 -

Table V: *Light and Heavy Naphtha Volume and Sulfur*

Case	LT Naphtha (vol%)	HY Naphtha (vol %)	LT Naphtha Sulfur (wppm)	HY Naphtha Sulfur (wppm)
ANS HVGO/HCGO Unhydrotreated	24.9	26.4	651	2009
ANS HVGO/HCGO Unhydrotreated	25.6	26.9	680	1920
ANS HVGO/HCGO Hydrotreated	34.3	26.6	5.1	15.7
ANS HVGO/HCGO Hydrotreated	31.7	27.7	30.0	80.0
100% Arabian HVGO Unhydrotreated	29.3	21.9	990	3907
100% Arabian HVGO Hydrotreated	29.1	24.8	3.0	26.0
80/20 Arabian HVGO/H-Oil Unhydrotreated	26.0	23.7	862	2945
80/20 Arabian HVGO/H-Oil Hydrotreated	41.4	21.6	10.0	59.7

430 fraction. This high sulfur level in this fraction could be attributed to benzothiophenes that were not converted in the FCC process. The only exception was the less severe hydrotreatment of the 80/20 HVGO/H-Oil® run which had 64.8% of the heavy naphtha sulfur. By undercutting the heavy naphtha, a significant decrease in the DB naphtha sulfur level could occur.

Table VI: Undercutting the Heavy Naphtha Portion
100% Arabian HVGO and 80/20 Arabian HVGO/H-Oil

Case	Temperature (F)	% Heavy Naphtha	Sulfur (wppm)	% Sulfur Contribution to 430 F Heavy Naphtha
100% HVGO Unhydrotreated	250 - 400	86.70	2850	63.3
	400 - 430	13.30	10800	36.7
	250 - 430	100.00	3907	
100% HVGO Unhydrotreated	250 - 400	86.84	2570	61.5
	400 - 430	13.16	10612	38.5
	250 - 430	100.00	3628	
100% HVGO Unhydrotreated	250 - 400	88.20	2720	57.4
	400 - 430	11.80	15100	42.6
	250 - 430	100.00	4181	
100% HVGO Unhydrotreated	250 - 400	88.37	2616	57.2
	400 - 430	11.67	14800	42.8
	250 - 430	100.00	4039	
80/20 HVGO/H-Oil Unhydrotreated	250 - 400	86.23	2030	59.4
	400 - 430	13.77	8674	40.6
	250 - 430	100.00	2945	
80/20 HVGO/H-Oil Unhydrotreated	250 - 400	85.85	1870	54.9
	400 - 430	14.15	9311	45.1
	250 - 430	100.00	2923	
100% HVGO Hydrotreated	250 - 400	84.85	17.4	57.3
	400 - 430	15.15	72.7	42.7
	250 - 430	100.00	25.8	
80/20 HVGO/H-Oil Hydrotreated	250 - 400	87.15	113	35.2
	400 - 430	12.85	1409	64.8
	250 - 430	100.00	280	
80/20 HVGO/H-Oil Hydrotreated	250 - 400	88.65	40	59.3
	400 - 430	11.35	214	40.7
	250 - 430	100.00	59.7	

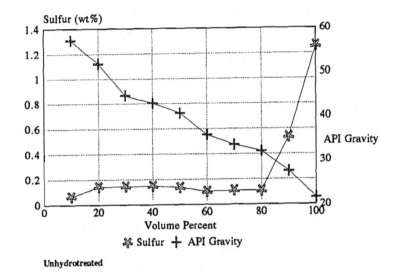

Figure 8: *Sulfur Distribution in Heavy Naphtha*

The sulfur distribution in the unhydrotreated feedstock is shown in Figure 8. The level of sulfur remained relatively constant near 1000 ppm until the 80 vol% point of the naphtha. The sulfur level increased dramatically to over 1.2 wt% in the last 10% of the naphtha fraction. This corresponded to an increase in aromatics in the back end of the heavy naphtha. Eliminating some of the heavy back end of the naphtha would help lower the sulfur level and the aromatic content of the gasoline pool, but an alternate use for this material would need to be developed.

CONCLUSIONS

1) Sulfur distribution in the FCC products is dependent on the initial feed sulfur, operating conditions of the FCC process, and the FCC feed source. If the FCC feed source is from a hydrotreater, the feed sulfur distribution will be very dependent on the operating conditions of the hydrotreater.

2) General hydrotreating benefits will include a higher FCC conversion with

an increase in the DB naphtha yield and a decrease in the coke and dry gas yield. The overall sulfur levels in all streams of the FCC will be reduced when comparing to an unhydrotreated feedstock.

3) Severe hydrotreating of the FCC feed can be done to comply with the new environmental regulations.

4) The relationship between the initial feed sulfur and the DB naphtha feed sulfur below 1500 ppm initial feed sulfur was linear. Beyond this point, the relationship becomes non-linear.

5) The linear to non-linear transition point in the initial feed sulfur-DB naphtha sulfur will vary with feedstock, FCC operating conditions, and hydrotreating operating conditions.

6) The T90 of the gasoline poll can be decreased in order to comply with the reformulated fuels requirements for lower gasoline sulfur since a large amount of the 430 EP gasoline is contained in the last 10% fraction, but an alternate use for this portion of the gasoline pool will need to be identified.

REFERENCES

1. NPRA Survey of U.S Gasoline Quality and U.S. Refining Industry Capacity to Produce Reformulated Gasolines: Part A. Jan. 1991.

2. NPRA Survey of U.S Gasoline Quality and U.S. Refining Industry Capacity to Produce Reformulated Gasolines: Part B. May 1991.

3. Wormsbecher, Richard F., Chin D.S., Gatte, R.R., Harding R.H., and Albro, T.G., "Catalytic Effects on the Sulfur Distribution in FCC Fuels," 1992 National Petroleum Refiners Meeting annual meeting at New Orleans, March 22-24, 1992.

4. Gailella, R.M, Andrews, J.W., Cosyns, J., and Heinrich, G., "Hydrotreating Around the FCCU," 1992 National Petroleum Refiners Meeting annual meeting at New Orleans, March 22-24, 1992.

22 Hydrocracking Phenanthrene and 1-Methyl Naphthalene: Development of Linear Free Energy Relationships

R. N. Landau[0], S. C. Korré, M. Neurock[1], M. T. Klein[2]

Center for Catalytic Science and Technology,
Department of Chemical Engineering,
University of Delaware, Newark, DE 19716

and

R. J. Quann

Mobil Research and Development Corporation
P.O. Box 480, Paulsboro, NJ 08066-0480

ABSTRACT

The catalytic hydrocracking reaction pathways, kinetics and mechanisms of 1-methyl naphthalene and phenanthrene were investigated in experiments at 350 °C and 68.1 atm H_2 partial pressure (190.6 atm total pressure), using a presulfided Ni/W on USY zeolite catalyst. 1-methyl naphthalene hydrocracking led to 2-methyl naphthalene, methyl tetralins, methyl decalins, pentyl benzene and tetralin. Phenanthrene hydrocracking led to dihydro, tetrahydro and octahydro phenanthrene, butyl naphthalene, tetralin to butyl

---

[0]Present Address: Merck Chemical Manufacturing Div., Rahway, New Jersey
[1]Present Address: Schuit Institute of Catalysis, Technical University of Eindhoven, The Netherlands
[2]Author to whom correspondence should be addressed

tetralin and dibutyl benzene. The rate constants for the dealkylation of butyl tetralins produced in the phenanthrene hydrocracking network conform to a linear free energy relationship (LFER), with the heat of formation of the leaving alkyl carbenium ion as the reactivity index.

INTRODUCTION

Catalytic hydrocracking is a versatile process for increasing the hydrogen to carbon ratio and decreasing the molecular weight of heavy oils. This versatility may prove extremely valuable in the search for optimal processing conditions and catalysts for production of "reformulated" gasolines. Associated reaction models will likely increase in detail as increasingly molecular output is desired. However, the complexity of hydrocracking feedstock structure and reactivity has kept traditional models somewhat global and, thus, often feedstock dependent. Therefore, the new, feedstock-sensitive, "molecular" models of hydrocracking reaction chemistry require the development of a critical mass of consistent molecular reaction pathways and kinetics as an essential data base. To this end, we report here on hydrocracking reaction pathways of 1-methyl naphthalene and phenanthrene, components among a broader set aimed at sampling the structural attributes of hydrocracking feedstocks. Special attention is devoted to the efficient organization of the resolved kinetic information into quantitative structure/reactivity correlations that will serve as a component of a broader kinetic data base.

The present work builds on the significant current understanding of the hydrocracking of paraffins (Froment, 1987) and extends to examine aromatic hydrocarbons. The aromatics' hydrocracking literature is less comprehensive. Several investigations on hydrocracking model bare ring polynuclear aromatic hydrocarbons involved Al_2O_3 or Si/Al_2O_3 catalysts (Qader, 1973; Shabtai et al., 1978; Lemberton and Guisnet, 1984). The effect of zeolite catalysts is examined in more recent publications (Haynes et al., 1983; Lapinas et al., 1987).

We report here on the reactions of 1-methyl naphthalene and phenanthrene. The experiments were focused on the target of discerning reaction families and characterizing them in terms of Quantitative Structure/Reactivity Correlations. We describe the work by first considering the experimental methods. The kinetics of 1-methyl naphthalene and phenanthrene hydrocracking are considered next. Finally, discernible linear free energy relationships are examined.

EXPERIMENTAL

Cyclohexane (HPLC grade, 99.99%, Aldrich) served as the solvent for the study of the reactions of the remaining compounds. Phenanthrene (98%+, Aldrich), served as the prototype three-ring aromatic moiety. Its staggered structure, makes it thermodynamically more stable than anthracene, and thus more abundant in heavy oils. 1-methyl naphthalene (98%, Aldrich) could in principle be obtained from the phenanthrene network, but its high rank in the network made parameters estimated statistically insignificant. All reactants were used as received.

The catalyst was a Mobil conditioned Zeolyst 753 Ni/W on USY zeolite, received in standard 3.0 mm pellets. Prior to all experiments, it was sulfided for two hours at 400° C by a 10% H_2S in H_2 gas stream (99%, Matheson Gas Products) at a flow rate of 30 cm^3/min. The catalyst was equilibrated by reaction with phenanthrene at 350 °C and 68.1 atm H_2 for _10 hours to achieve a steady-state activity that lasted about 50 more hours on stream.

A one-liter spinning basket batch autoclave (modified from original as received from Autoclave Engineers) was the core of the reaction system. A detailed description of the system is available elsewhere (Landau, 1991). The catalyst basket was mounted on the autoclave's agitator which was equipped with baffles to ensure turbulence. Varying the stirring speed revealed the absence of diffusion limitations for a stirring rate of 10 s^{-1} or greater. Also, phenanthrene hydrocracking with powdered catalyst (50-200 mesh) in a slurry exhibited the same kinetic behavior as with the 3.175mm pellets, which suggested that internal transport limitations were not important.

Experiments were routinely performed at 350°C and 68.1 atm H_2 (190.6 atm total pressure, the balance from cyclohexane vapor pressure), with 10g of catalyst in the spinning basket arrangement. The stirring rate was 15 s^{-1}, increased by 50% for 10 min. immediately following the injection. Total pressure was regulated with continuous hydrogen makeup. Sampling was scheduled to remove no more than 5% of the total liquid volume.

Products were identified using a Hewlett Packard 5970 Mass Selective Detector and a Hewlett Packard 5880A Gas Chromatograph, employing a fused silica capillary column and flame ionization detector. Dibenzyl ether (Aldrich, > 99.9%) was used as an external standard.

RESULTS

Hydrocracking of 1-methyl naphthalene

Reaction of 25% wt 1-methyl naphthalene took place in cyclohexane as a solvent at 68.1 bar H_2 at 350 °C, and led to 6 identified isomeric lumps, representing 99%+ material balance closure. Kinetics are summarized in Fig. 1. There are two primary products of 1-methyl naphthalene hydrocracking, both with high selectivity : methyl tetralins (lump including 1, 2, 5, and 6-methyl tetralins) and 2-methyl naphthalene (isomerization product). The balance consists of methyl decalins, pentyl benzenes and tetralins.

Delplot analysis classifies pentyl benzenes as secondary products (direct cracking of methyl tetralins), as well as methyl decalins. Tetralin appeared to be weakly primary product, which implies that their formation is the result of disproportionation reactions of methyl tetralins with the solvent, since no bare-ring naphthalene has been observed.

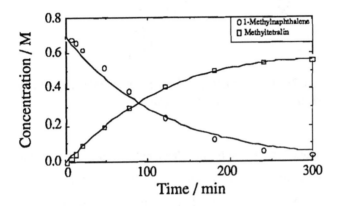

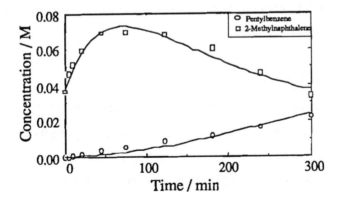

Figure 1: *Temporal yields of some 1-methyl naphthalene hydrocracking products. Solid lines represent parameter estimation results.*

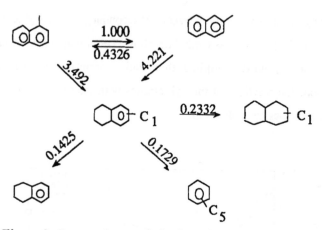

Figure 2: *Proposed network for 1-methyl naphthalene hydrocracking (normalized with 1.49 10^{-3} l/g_{cat}/min.)*

Parameter estimation for the above network provided more insight into the pathways. Isomerization of 1-methyl naphthalene occurred at a slower rate than its hydrogenation, while 2-methyl naphthalene hydrogenated at a slightly higher rate than 1-methyl naphthalene. Further reactions of methyl tetralins (hydrogenation to decalins, cracking to pentyl benzenes) occurred with rate parameters one order of magnitude lower.

Hydrocracking of Phenanthrene

Reaction of 3.4 wt% phenanthrene in cyclohexane solvent and at 68.1 atm H_2 at 350 °C led to 11 identified isomeric product lumps, representing 95%+ material balance closure. Each lump represents grouping of molecular weight isomers. For example the tetrahydro phenanthrenes lump includes molecules with cyclohexyl and methyl cyclopentyl saturated rings. *Sym-* and *asym-* octahydro phenanthrenes are in the same lump, and "butyl tetralin" refers to any tetralinic unit sheet with a side chain of four carbon atoms, regardless of its position on the unit sheet.

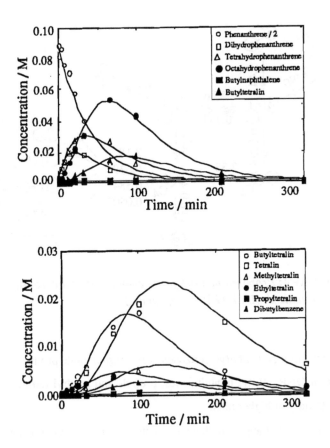

Figure 3: *Temporal yields of some phenanthrene hydrocracking products. Solid lines represent model correlations.*

Reaction kinetics are summarized in Fig. 3, which shows, by the initial positive slopes, that dihydrophenanthrene (DHP) and tetrahydro phenanthrenes (THP) were the primary products, DHP forming with higher initial selectivity. Octahydro phenanthrenes (OHP) had very low initial selectivity and were interpreted as secondary products, evolving from THP. Butyl tetralins, ethyl

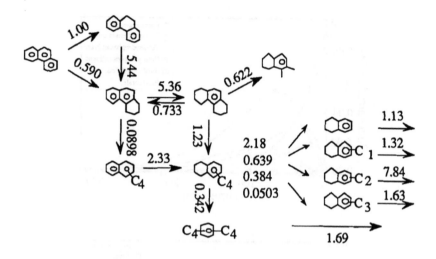

Figure 4: *Proposed network for phenanthrene hydrocracking (normalized with 1.49 10⁻³ l/gcat/min.)*

tetralins, tetralins and dibutyl benzenes were higher rank products, evolving mainly from OHP. Butyl naphthalenes, propyl- and methyl tetralins were all clearly of tertiary or higher rank.

DHP did not appear to dehydrogenate back to phenanthrene, as reported over non-acidic catalyst supports (Girgis and Gates, 1991), but rather hydrogenated further to THP. This difference was probably due to the higher H2 partial pressure employed in this work. THP in turn underwent further hydrogenation to OHP; this was more selective than ring opening to n-butyl naphthalene. The terminal naphthenic ring in OHP was cracked to butyl tetralin, and cleavage of the butyl side chain occurred at various positions.

The optimized kinetic parameter fitting on all the network components of Fig. 4 is presented in Fig. 3, as modeled species' concentration vs. time. Clearly the fit is good. The chemical significance of the rate constants regressed is thus the remaining issue. This is considered more fully below.

LFER DEVELOPMENT

The organization and chemical significance of these kinetics data can be enhanced by the existence of linear free energy relationships (LFER). A LFER will exist for a reaction family with similar transition state sterics (essentially constant A-factor) and reactivities that differ because of differences in activation energies. This will assume the form of a linear correlation between the reactivity of a molecule, as it is expressed through its rate parameter, and its structure, as expressed by a reactivity index pertinent to each reaction. The existence of a LFER not only helps to establish the reaction mechanism, but can also concisely summarize an enormous amount of information in a handful of slopes and intercepts. The data bases constructed in this way will be general and flexible enough to draw useful correlations for process modeling.

The applicability of LFERs in heterogeneous catalysis has been hindered in part because of the uncertainty of controlling elementary steps (Dunn, 1968). Observable kinetics are generally the expression of many elementary steps acting in concert. However, Mochida and Yoneda (1967) uncovered a linear relationship between the logarithm of the observed rate constant for dealkylation of a particular alkyl benzene and the enthalpy change for hydride abstraction from the related paraffin, $DH_C^+(R)$ (Olah et al., 1964). The mechanistic information in this correlation is the suggestion of a rate limiting step for the dealkylation reactions that correlates with the formation of the alkyl carbenium ion. It is also possible that several steps correlate in concert with the same reaction family index. In any case, the modeling value of this correlation is that it allows for the *a priori* prediction of other dealkylation reactions. We follow this perspective in the search for useful correlations to summarize hydrocracking kinetics data.

The butyl tetralin dealkylation reactions observed during hydrocracking of phenanthrene seemed a reasonable point to search for a

found by Mochida and Yoneda (1967). Careful scrutiny of the kinetic parameters regressed reveals a trend for the dealkylation of butyl tetralins. The dealkylation rate constants increase as the stability of the leaving alkyl carbenium ion increases. This is consistent with the energetics of the formation of the alkyl carbenium ions. This suggests that the formation of the alkyl carbenium ion could contribute to the controlling energetics of the process.

This information was tested more quantitatively using a linear free energy relationship for the butyl tetralin dealkylation reactions with the stability of the dealkylating carbenium ion as the reactivity index. Two options for the value of the reactivity index were available, reflecting the nature of the alkyl carbenium ion: the energetics of either the primary (n-alkyl) or the most stable alkyl carbenium ion could be used. The experimental data were not sufficiently precise to adjudicate.

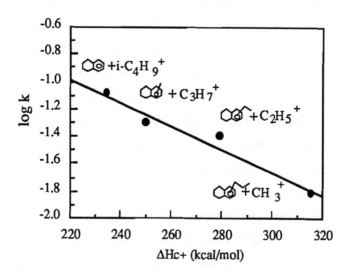

Figure 5: *LFER for the butyl tetralin dealkylation reactions. (k in min.$^{-1}$)*

CONCLUSIONS

1-methyl naphthalene hydrocracking resulted in isomerization to 2-methyl naphthalene, as well as hydrogenation to methyl tetralins. Hydrogenation of the single ring in methyl tetralin was one order of magnitude slower than hydrogenation of methyl naphthalenes, and comparable to the naphthenic ring opening reactions. Phenanthrene hydrocracking proceeded mainly through sequential hydrogenation to dihydro, tetrahydro and octahydro phenanthrene, followed by opening of the terminal naphthenic rings and dealkylation of the butyl side chains.

Reasonable structure/reactivity correlations were obtained for the rate parameters of the dealkylation of butyl tetralins produced from phenanthrene hydrocracking. The heat of formation of the leaving alkyl carbenium ion $(CH_3^+\text{-}C_5H_{11}^+)$ as used from Mochida and Yoneda (1967) was established as a suitable reactivity index.

ACKNOWLEDGEMENT

The authors acknowledge the financial support of Mobil Research and Development Corporation (Paulsboro Research Laboratory), and the State of Delaware, as authorized by the State Budget Act of Fiscal Year 1989.

REFERENCES

1. Bhore, N. A., Klein, M.T., and Bischoff, K.B. (1990). The Delplot Technique: A New Method for Reaction Pathway Analysis. *Ind. Eng. Chem.*, 29(2), 313-316.
2. Dunn, I.J. (1968). Linear Free Energy Relationships in Modeling Heterogeneous Catalytic Reactions. *J. Catal.*, 12, 335-340.
3. Froment, G.F. (1987). Kinetics of the Hydroisomerization and Hydrocracking of Paraffins on a Platinum Containing

Bifunctional Y-Zeolite. *Catal. Today*, <u>1</u>, 455-473.

4. Girgis, M.J., and Gates, B.C. (1991). Reactivities, Reaction Networks, and Kinetics in High-Pressure Catalytic Hydroprocessing. *Ind. Eng. Chem. Res.*, <u>30</u>(9), 2021-2058.

5. Haynes, H.W.J., Parcher, J.F., and Helmer, N.E. (1983). Hydrocracking Polycyclic Hydrocarbons over a Dual-Functional Zeolite (Faujacite)-Based Catalyst. *Ind. Eng. Chem. Process Des. Dev.*, <u>22</u>, 401-409.

6. Landau, R.N. (1991) <u>Chemical Modeling of the Hydroprocessing of Heavy Oil Feedstocks</u>. Ph.D. Thesis, University of Delaware.

7. Lapinas, A.T., Klein, M.T., Gates, B.C., Macris, A., and Lyons, J.E. (1987). Catalytic Hydrogenation and Hydrocracking of Fluoranthene: Reaction Pathways and Kinetics. *Ind. Eng. Chem. Res.*, <u>26</u>, 1026-1033.

8. Lemberton, J.-L., and Guisnet, M. (1984). Phenanthrene Hydroconversion as a Potential Test Reaction for the Hydrogenating and Cracking Properties of Coal Hydroliquefaction Catalysts. *Appl. Catal.*, <u>13</u>, 181-192.

9. Mochida, I., and Yoneda, Y. (1967). Linear Free Energy Relationships in Heterogeneous Catalysis I. Dealkylation of Alkylbenzenes on Cracking Catalysts. *J. Catal.*, <u>9</u>, 386-392.

10. Olah, G.A., Baker, E.B., Evans, J.C., Tolgyesi, W.S., McIntyre, J.S., and Bastien, I.J. (1964). Stable Carbonium Ions. V. Alkylcarbonium Hexafluoroantimonates. *J. Am. Chem. Soc.*, <u>86</u>(1360-1372).

11. Pines, H. (1981). <u>The Chemistry of Catalytic Hydrocarbon Conversions</u>. New York, N. Y.: Academic Press.

12. Qader, S. A. (1973). Hydrocracking of Polynuclear Aromatic Hydrocarbons over Silica-Alumina Based Dual Functional Catalysts. *J. Inst. Pet.*, <u>59</u>, 178-187.

13. Shabtai, J., Veluswami, L., and Oblad, A.G. (1978). Steric Effects in Phenanthrene and Pyrene Hydrogenation Catalyzed by Sulfided NiW/Al$_2$O$_3$. *Am. Chem. Soc., Div. Fuel Chem. Prepr.*, <u>23</u>(1), 107-112.

23 A Novel Process for Upgrading Heavy Oil Emulsions

F.T.T. Ng and R.T. Rintjema

Department of Chemical Engineering, University of Waterloo, 200 University Avenue West, Waterloo, Ontario, Canada N2L 3G1

ABSTRACT

Canada has extensive reserves of high sulphur heavy oils. These heavy oils are recovered primarily by steam injection techniques. As a result, the heavy oils are obtained as emulsions at well-heads. The heavy oils, being high in sulphur and metals, and low in hydrogen to carbon atomic ratio, require upgrading such as desulphurization and hydrocracking before it can be used in conventional refineries. Conventional emulsion treatment and desulphurization technology require multistage processing. Thus, alternative technologies for processing heavy oil emulsions would be attractive.

We have recently developed a novel single stage process for upgrading emulsions via activation of water to provide hydrogen *in situ* for catalytic desulphurization and hydrocracking. Current work is focused on the desulphurization aspect of upgrading, using benzothiophene as the model sulphur compound and molybdic acid as the catalyst. At 340°C and a CO loading pressure of 600 psi, up to 86 % sulphur removal was obtained. As well, *in situ* generated H_2 was found to be more active than externally supplied molecular H_2. A likely pathway for desulphurization of benzothiophene was via the initial hydrogenation of benzothiophene to dihydrobenzothiophene followed by hydrogenolysis to give ethylbenzene and H_2S.

INTRODUCTION

As conventional oil reserves continue to decrease it will become necessary to utilize high sulphur (sour) heavy crude oil or bottoms in the near future. This is particularly true for Canada which has extensive reserves of heavy crude oils. Current technologies for the recovery of heavy oils involve steam flooding, cyclic steam injection, or fire flooding of reservoirs. These methods often produce stable water-in-oil emulsions which must be treated prior to being transported by pipeline. In addition, Canadian heavy oils typically contain 2 to 6 wt% sulphur [1]. A significant amount of this sulphur must be removed because of the potential for catalyst poisoning as well as environmental considerations. Therefore, emulsion treatment and desulphurization of heavy oils must occur before recovered heavy oils can be utilized in a refinery process.

Conventional upgrading techniques for heavy oils include visbreaking to reduce viscosity, as well as direct hydrogenation using molecular hydrogen at high temperatures and pressures (up to 425°C and 2500 psi [2]). Complete separation of the water from the oil requires costly physical separation devices, such as induced air flotation, or the addition of costly specialty chemicals, such as demulsifiers. The economics of the recovery and upgrading of heavy oils is the dominant drawback to fully utilizing these reserves. The cost of equipment and chemicals, as well as the cost of supplying, storing, and maintaining high purity hydrogen can be very unattractive.

Our laboratory has recently developed a novel single stage process for upgrading emulsions via activation of water to provide hydrogen *in situ* through the water gas shift reaction (WGSR) for catalytic desulphurization [3]. This process achieved complete emulsion breaking and high sulphur removal, without the need for any prior emulsion treatment or externally supplied molecular hydrogen. Furthermore, reaction conditions were well within the operating range of industrial hydrotreaters. Our current work has been focused on one aspect of upgrading, namely desulphurization. The upgrading procedure was achieved using the water soluble catalyst molybdic acid (MA) and a water-in-toluene (H_2O/TOL) emulsion. Benzothiophene was chosen as a model sulphur containing compound for desulphurization studies.

EXPERIMENTAL

All experiments were performed in a 300 ml, 316 stainless steel batch autoclave from Autoclave Engineers. Once the emulsion system was placed in the reactor, an involved sealing procedure was undertaken to ensure no leaks were present. For all experiments, the temperature of the reactor was increased at a rate of 2°C per minute as recommended by Autoclave Engineers. At the end of the reaction period the reactor was allowed to cool naturally overnight to room temperature. Gas, liquid, and solids were then collected and analyzed. The gas product was collected in a gas sample bag and was immediately analyzed using gas chromatography. Gas analysis was performed on a Perkin-Elmer Model 8500 gas chromatograph equipped with a thermal conductivity detector. The inlet system was composed of a 1.5 m Hayesep C column in series with a 2 m molecular sieve column. Liquid analysis was carried out using the Perkin-Elmer Model 8500 gas chromatograph equipped with a 30 m DB-1701 fused

Table 1: *Standard Emulsion Reactant System*

Component	ml	g
Toluene (TOL)	49	42.5
Benzothiophene (BTH)	9.8	11.3[1]
Deionized Water	21	21
Emulsifier (BASF P105)	~0.4	0.7[2]
Molybdic Acid (MA)	NA	0.5324[3]
Total	80.2	75.5

[1] Equivalent to 80 mmol BTH.
[2] Equivalent to approximately 0.9 wt % emulsifier.
[3] Equivalent to 4775 wppm Mo.

silica capillary column and a flame ionization detector. A VG Trio-1S bench top GC/MS was used for identification of liquid products. X-ray fluorescence Oxford model Lab X-1000 was used to determine Mo and S content in the solid product obtained. The components of the "standard benzothiophene emulsion" are given in Table 1.

Preparation of the emulsion was as follows. The BTH was first dissolved in the toluene with the emulsifying agent BASF P105. The MA catalyst was dissolved in the deionized water and then added to the toluene phase and shaken vigorously to form a stable water-in-oil emulsion. The emulsion was then transferred to a 316 stainless steel liner and processed immediately.

RESULTS AND DISCUSSION

Catalytic desulphurization (HDS) of benzothiophene (BTH) was achieved by activation of H_2O to generate H_2 *in situ* via the water gas shift reaction (WGSR), according to reaction (1).

$$CO + H_2O \rightleftharpoons CO_2 + H_2 \tag{1}$$

The generated *in situ* H_2 was subsequently utilized for the HDS of BTH. The stoichiometry of the HDS reaction to yield ethylbenzene (EB) is given by reaction (2).

$$\text{BTH} + 3\,H_2 \longrightarrow \text{EB-}C_2H_5 + H_2S \tag{2}$$

The catalytic desulphurization of BTH was carried out under an initial CO loading pressure (P_{coi}) of 600 psi. Prior to reaching the reaction temperature of 340°C, essentially no conversion of BTH was observed. System pressure increases during the heat up period and typically reaches 2950 psi at 340°C. Following an induction period of 1 h, the total pressure of the system will decrease by approximately 200 to 300 psi. This decrease in pressure can be attributed to the stoichiometry of the HDS reaction (reaction (2)), whereby 3 moles of H_2 react with 1 mole of BTH to give 1 mole of EB and 1 mole of H_2S. At the end of the reaction the emulsion was completely broken into a distinct water and a distinct toluene phase. Fine black solids are seen to accumulate near the water/toluene interface, which are found to be some form of a molybdenum sulphide species. Gas phase products were found to include H_2, CO_2, CO, and H_2S. The liquid products were mainly EB with minor amounts of dihydroxybenzothiophene (DHBT).

Mechanisms for the desulphurization of BTH have been reviewed by a number of researchers [4-7]. Two reaction schemes have been proposed. One involves the initial hydrogenation of BTH to DHBT, followed by sulphur removal to give EB and H_2S. The second scheme is direct desulphurization via C-S bond cleavage resulting in H_2S and styrene (STY). The STY intermediate

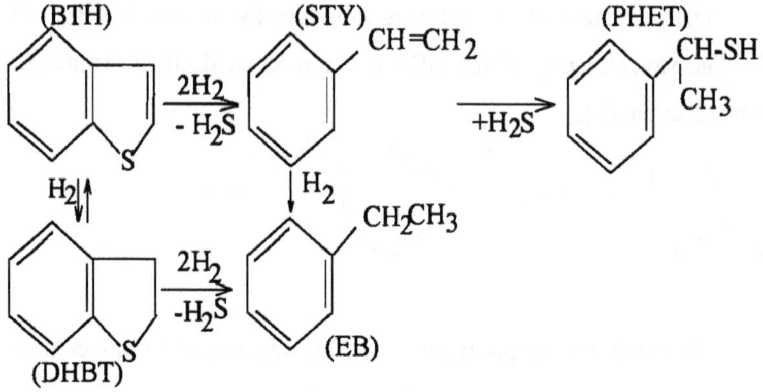

Figure 1: *Proposed reaction mechanism for HDS of BTH*

then either reacts with H_2S to yield phenylethanethiol (PHET) or is hydrogenated directly to EB. These reaction pathways, as proposed by Guin et al. [4], are shown in Figure 1.

Since DHBT was detected in the product, and no STY or PHET were detected, a likely route for desulphurization is via the DHBT intermediate pathway. This pathway has also been suggested for the desulphurization of BTH by phosphomolybdic acid [8] and molybdenum naphthenate [9]. It is interesting to note that the DHBT pathway is also supported by thermodynamics, as the initial hydrogenation of BTH to DHBT is more favourable than direct desulphurization to STY and H_2S [9].

From the reaction schemes presented by reactions (1) and (2), it is clear that the degree of HDS of BTH attained can be measured by the amount of EB produced since the DHBT molecule still contains sulphur. The extent to which the WGSR and the HDS reaction proceed are expressed in terms of the following variables: CO conversion, BTH conversion, and sulphur removal. CO and BTH conversions are based on the initial and final mole amounts of each component. Calculations for BTH conversion have been normalized to take into account minor losses of the liquid phase that occur during product collection. Sulphur removal was calculated by dividing the moles of EB formed by the initial moles of BTH present (BTH_i).

The catalytic activity of MA for the HDS of BTH through *in situ* H_2 generation at 340°C is demonstrated by the data in Table 2. The two experiments conducted without MA and the experiment run under a N_2 atmosphere were designed to determine the conversion of CO and BTH due to the inherent reactivity of the reactor system and the thermal conversion of BTH at 340°C. Results indicated that without MA a 33 - 38 % CO

Table 2: *Activity of MA for the HDS of BTH at 340°C[1]*

	WGSR	3 h HDS	5 h HDS	No MA WGSR	No MA WGSR	No CO
Conditions						
loading gas	CO	CO	CO	CO	CO	N_2
mmol gas	300	300	300	300	300	300
BTH (mmol)	0	80	80	0	80	80
Reaction Time (h)	8.0	3.0	5.0	3	3	3
Results (mmol)						
H_2	109	32	8	56	47	0
CO_2	137	145	113	67	68	0
CO	62	51	40	189	193	0
H_2S	--	24	30	--	0	0
BTH	--	19	17	--	71	76
EB	--	48.5	60.5	--	5	2
DHBT	--	12.5	2.5	--	4	2
% Conversion						
CO	80	82	86.5	38	33	--
BTH	NA	76	78.5	--	11	7
S removal[2]	NA	61	76	--	6	3

[1] 4775 wppm Mo as MA; P_{COi} = 600 psig; TOL/H_2O = 2.3 ml/ml; 550 rpm stir speed

conversion was obtained. The thermal conversion of BTH was about 10 %, with a 3-6 % sulphur removal. In the absence of BTH, but under a CO loading pressure of 600 psig, a reaction temperature of 340°C and a MA concentration of 4775 wppm Mo-metal, the WGSR produces a CO conversion of 80 % over a 8 h reaction time. With the addition of 80 mmol of BTH, 82.5 % CO conversion was reached in 3 h. Over a 5 h reaction time, however, the CO conversion was increased even further to 86.5 %. It appears that the consumption of H_2, as required by the HDS reaction, shifts the equilibrium of the WGSR towards the right as expected according to equations (1) and (2). The quantity of H_2 produced, as well as CO conversion, increase greatly when MA was present in the system. Therefore, MA catalyzes the WGSR.

The results for the HDS experiments illustrate that 76 % BTH conversion can occur in 3 h (at 340°C and under an initial CO loading pressure of 600 psig) and increases only marginally to 78.5 % when the reaction time was increased to 5 h. However, the product distribution for these two experiments are different. In 3 h, a sulphur removal of 61 % was reached and a significant quantity of DHBT intermediate was observed. At longer reaction times (5 h), sulphur removal increases significantly to 76 % over 5 h. It should be mentioned that the equilibrium BTH conversion value of 97 % [10] was not approached in 5 h of reaction time. MA also catalyzes the HDS of BTH since the quantity of EB produced, as well as BTH conversion and sulphur removal, increase dramatically in the presence of MA.

In industrial practice, the WGSR can be described as a two-step, two-catalyst, heterogeneous process [11]. Most industrial feedstocks contain a number of sulphur compounds which invariably poison most heterogeneous shift catalysts. The large values of CO conversion obtained in the presence of BTH (Table 2) indicates that MA is an effective sulphur-tolerant WGSR catalyst. Results of the desulphurization of BTH under a 600 psig loading pressure at 340°C with varying initial concentrations of BTH, are shown in Table 3. A gradual increase in CO conversion can be observed with increases in BTH concentration, reaching a maximum at 86.5 % for 80 mmol of BTH. This suggests that the presence of larger amounts of BTH causes a greater shift in the WGSR to the right because of the consumption of greater amounts of H_2. The conversion of BTH does not follow the same trend as CO conversion. A maximum value of BTH conversion of 88 % was reached for 60 mmol BTH

Table 3: *Effect of BTH loading on the HDS of BTH at 340°C[1]*

Conditions				
BTH (mmol)	5	40	60	80
Reaction time (h)	6.8	5.0	5.0	5.0
Results (mmol)				
H_2	110	60	26.5	8
CO_2	133	146.5	145	113
CO	72.5	57	56	40
H_2S	0	15.5	21.5	30
BTH	1.2	5.5	6.5	17
EB	3.8	34.5	49.5	60.5
DHBT	0	0	4	2.5
% CO conversion	76	81	82	86.5
% BTH conversion	76.5	86	88	78.5
% S removal[2]	76.5	86	83	76

[1] 4775 wppm Mo as MA; P_{CO} = 600 psig; TOL/H_2O = 2.3 ml/ml; 550 rpm stir speed
[2] (mmol EB/mol BTH) x 100

before dropping to 78.5 % at the 80 mmol BTH level. Sulphur removal follows a similar path as BTH conversion with a maximum of 86 % removal for 40 mmol of BTH, and dropping to 76 % for 80 mmol BTH. Furthermore, no DHBT was detected until 60 mmol of BTH was employed. With 80 mmol of BTH, the percentage of the product that appeared as DHBT decreases slightly over that for the 60 mmol experiment.

The decrease in BTH conversion at the 80 mmol level can be explained by the fact that the amount of H_2 available was no longer in excess. For a 600 psig loading of CO (300 mmol) and an 86.5 % CO conversion, 260 mmol of H_2 are produced theoretically. Therefore, the maximum quantity of BTH that can be converted to EB is 86 mmol, based on the stoichiometry of equation (2).

To determine the effect of the amount of catalyst used for the reaction process, the MA concentration was varied from 0 to 4775 wppm Mo-metal. The results of these experiments are shown in Table 4.

The only significant changes in conversions and sulphur removal occur at MA concentrations below 2400 wppm Mo-metal. It appears that the reaction system becomes saturated with catalyst in the 2400-4775 wppm Mo-metal region.

Table 4: *Effect of MA concentration on the WGSR and HDS of BTH at 340 °C[1]*

Mo^2 wppm	% CO conversion	% BTH conversion	% S removal
0	33	11	6
400	74	57	38
1200	81	76	55
2400	85	82	76
4775	82	76	61

[1] P_{CO} = 600 psig; 80 mmol BTH, 3 h reaction time; TOL/H_2O = 2.3 ml/ml; 550 rpm stir speed.
[2] Expressed as wppm Mo in MA.

The effect of H_2O/toluene ratio on the BTH emulsion upgrading process was carried out with experiments conducted at H_2O/CO ratios of 1.0, 4.0, and 6.7 (mol/mol). At the end of the reaction, the emulsion was found to be broken into a distinct water and toluene phase in all cases. The results in Table 5 indicate that increasing the H_2O/CO ratio from 1.0 to 6.7 produces no significant change in CO conversion. As well, no noticeable difference exists in BTH conversion or sulphur removal when the H_2O/CO molar ratio is 1.0 or 4.0. However, BTH conversion and sulphur removal decrease significantly when the H_2O/CO molar ratio was raised to 6.7. Therefore, large concentrations of H_2O appear to impede the HDS reaction process. It should be noted that the MA concentration for the experiment with a H_2O/CO molar ratio of 1.0 was less than that for the two higher H_2O concentration

Table 5: *Effect of H_2O content on the HDS of BTH at 340 °C[1]*

Conditions			
H_2O (mmol)	300^3	1200	2000
H_2O/CO (mol/mol)	1.0	4.0	6.7
TOL/H_2O (ml/ml)	8.2	2.3	1.0
Results (mmol)			
H_2	30	32	65
CO_2	140.5	145	168.5
CO	68	24	51
H_2S	21.5	51	13
BTH	18	19	29
EB	50	48.5	34
DHBT	12	12.5	17
% CO conversion	79	82	82
% BTH conversion	77.5	76	64
% S removal[2]	62.5	61	42.5

[1] 4775 wppm Mo as MA; P_{CO} = 600 psig; 80 mmol BTH, 3 h reaction
[2] time; TOL/H_2O=2.3 ml/ml; 550 rpm stir speed.
[3] (mol EB/mol BTH$_i$) x 100.
    3000 wppm Mo as MA used in this experiment.

experiments. This was due to the limited solubility of MA in H_2O. However, the results shown in Table 4 indicate that no significant difference in the conversions will be obtained by operating at 3000 wppm Mo rather than at 4775 wppm Mo. Therefore, the effect of changing H_2O/CO ratio on BTH conversion and sulphur removal may be related to the competition between H_2O and BTH for active sites, or changes in emulsion properties and gas solubilities because of different amounts of H_2O.

An experiment was performed to compare the activity of *in situ* generated H_2 with externally supplied molecular H_2. The results are shown in Table 6. At 340°C and a CO loading pressure of 600 psig (300 mmol), 86.5 % CO conversion was achieved in 5 h for the *in situ* H_2 (Table 3). This corresponds to the production of 260 mmol H_2 according to the stoichiometry of the WGSR given in reaction (1), which is equivalent to a loading pressure of 550 psig H_2 at 25°C.

The HDS experiment with externally supplied molecular H_2 achieved

Table 6: *Activity of in situ H_2 versus externally supplied H_2 at 340°C[1]*

Conditions	supplied H_2	in situ H_2
loading gas	H_2	CO
pressure (psig)	550	600
mmol gas	260	300
reaction time (h)	7.0	5.0
Results (mmol)		
BTH	27.5	17
EB	41	60.5
DHBT	11.5	2.5
% BTH conversion	65.5	78.5
% S removal[2]	51	76

1   4775 wppm Mo as MA; 80 mmol BTH, TO = L/H_2O = 2.3 ml/ml; 550 rpm stir speed.
2   (mol EB/mol BTH$_i$) x 100.

a BTH conversion of 65.5 % and a sulphur removal of 51 %. Although this experiment was conducted for 2 h longer than the *in situ* generated H_2 experiment, BTH conversion and sulphur removal values are significantly lower. This finding suggested that *in situ* produced H_2 was more active than externally supplied H_2 for BTH conversion and hydrogenolysis (sulphur removal). Activity toward the HDS reaction could be related to the ease with which a metal-hydride active species could be formed. Hook and Akgerman [12] also found *in situ* generated H_2 to be more active than externally supplied H_2 for the HDS of dibenzothiophene. These researchers attributed the greater activity to the formation of hydrogen at the surface, where it could resemble atomic hydrogen or a form of metal hydride.

CONCLUSIONS

Molybdic acid was found to be an effective sulphur-tolerant catalyst for the WGSR and the HDS of BTH at 340°C. The conversion of BTH can be as high as 88 %, and sulphur removal as high as 86 %. With the addition of 80 mmol of BTH, CO conversion was increased by 6.5 %. The increase in CO conversion was related to the consumption of large amounts of H_2 by BTH, thus shifting the WGSR further to the right to increase production of H_2. Based on BTH conversion and sulphur removal, *in situ* generated H_2 was more active than externally supplied molecular H_2. A likely pathway for desulphurization was via the initial hydrogenation of BTH to DHBT followed by hydrogenolysis to give EB and H_2S.

ACKNOWLEDGEMENTS

Financial assistance from the Natural Sciences and Engineering Research

Council of Canada, Strategic Grant Program, is gratefully acknowledged.

REFERENCES

1)   Clark, P.D., Clarke, R.A., Hyne, J.B., Lesage, K.L., *AOSTRA J. Res.*, **6**, 29-39 (1990).

2)   Shell Group of Companies, The Petroleum Handbook 6ᵗʰ ed., Elsevier, Amsterdam, 1983, 307-313.

3)   Ng, F.T.T., Rintjema, R.T., Studies on Surface Science and Catalysis **73**, Smith, K.J.,Sanford, E.C. (Editors) Elsevier, 51-58 (1992).

4)   Guin, J.A., Lee, J.M., Fan, C.W., Curtis, C.W., Lloyd, C.W., Tarrer, A.R., *Ind. Eng. Chem. Process Des. Dev.*, **19**, 440-446 (1980).

5)   Daly, F.P., *J. Catal.*, **51**, 221-228 (1978).

6)   Furimsky, E., Amberg, C.H., *Can. J. Chem.*, **54**, 1507-1511 (1976).

7)   Choi, M.G., Robertson, M.J., Angelici, R.J.,*J. Am. Chem. Soc.*, **113**,   4005-4006 (1991).

8)   Ng, F.T.T. and Tsakiri, S.K., *Fuel*, **71**, 1309-1314 (1991).

9)   Ng, F.T.T. and Walker, G.R., *Can. J. Chem. Eng*, **69**, 844-851 (1991).

10)   Rintjema, R.T., *MASc. Thesis*, University of Waterloo (1992).

11)   Laine, R.M., Crawford, E.J., *Mol. Catal.*, **44**, 357-387 (1988).

12)   Hook, B.D., Akgerman, A., *Ind. Eng. Chem. Process Des. Dev.*, **25**, 278-284 (1986).

24 Hydroconversion of Methyl-Cyclohexane on a Bifunctional Catalyst

S. MIGNARD, Ph. CAILLETTE and N. MARCHAL
Institut Français du Pétrole, B.P. 311, 1-4 avenue de Bois Préau,
92506 Rueil-Malmaison, France

ABSTRACT

The hydroconversion of methylcyclohexane has been studied on a Pt/USY catalyst under hydrocracking conditions. Methylcyclohexane is firstly isomerized in dimethylcyclopentanes and then is transformed in C7 alcanes by a first C-C bond breakage. A second C-C bond breakage of the C7 alcanes yields to cracked products C3 and C4. It is shown that the second C-C bond breakage occurs very rapidly and so, the C7 alcanes cannot be obtained with a high selectivity. The distribution of the C7 products shows that all the isomers are at the thermodynamic equilibrium.

INTRODUCTION

Gas oils produced from catalytic cracking are rich in monoaromatic (10-15 wt%), diaromatics (40-50 wt%) and triaromatic (5-15 wt%) components which have very low cetane values. Hydrogenation of these compounds would permit a substantial improvement in the cetane value of the gas oil, for example the hydrogenation of naphthalene (cetane=0) to decaline (cetane=48). However, it is theoretically possible to further increase the cetane value by converting the hydrogenated cyclic alkanes to non-cyclic alkanes. The conversion of decylcyclohexane to a non-cyclic iso-C16 then to n-C16 would permit a gain in the cetane value from 40 to 55 then to 100.

Although ring opening of a cyclo-alkane could significantly increase the quality of gas oil fractions, cracking of the resulting alkanes must be minimized in order to maintain the overall yield of gas oil. The ring opening reaction has been studied under hydrocracking conditions using a typical bifunctional catalyst, see for example [1-4]. In this work, the reaction was carried out on methylcyclohexane under high hydrogen partial pressure and moderate temperature.

EXPERIMENTAL

Catalyst Preparation. A Pt/USY catalyst was prepared and used for catalytic testing. Platinum deposition was carried out by competitive ion exchange using $Pt(NH_3)_4Cl_2$ and NH_4OH as the competitor. A Pt/NH_4^+ mole ratio of 50 was employed. The ratio of volume of solution to weight of USY was 8l/kg. The Y`zeolite chosen for this study had the formula $H_1(AlO_2)_1(SiO_2)_{20} \cdot xH_2O$. The dried and calcined catalyst contained 0.65 wt% Pt.

Operating Conditions. Experiments were performed at a total pressure of 60 atm. The contact time and temperature were varied from 0.33 to 4 h and from 513K to 633K respectively. Testing was performed on <u>in situ</u> calcined (793K, 2h) and reduced (723K, 2h) catalysts.

The reactant/products have been classified as follows:

MC :   unreacted methylcyclohexane
I :   methylcycloalkane isomerisation products, i.e. dimethylcyclopentanes and ethyl-cyclopentane
RO :   ring opened C_7 alkane isomerisation products
C :   cracked products, i.e. products resulting from a second C-C bond cleavage reaction on the RO products.

RESULTS

Influence of the reaction temperature. The variation of the effluent composition as a function of the reaction temperature is shown in Fig.1. The percentage of unconverted MC decreases nearly linearly between 513K and 633K. The product distribution is characterized by three zones: Z1 between 513K and 553K where only the I (isomerisation) products appear, Z2 between

553K and 593K where the RO and C products begin to appear at the expense of the I products and Z3 above 593K where the C products are formed in appreciable quantity.

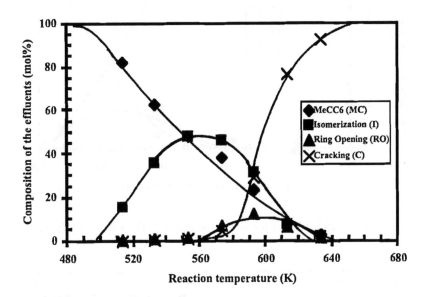

Figure 1 : *composition of the effluents vs temp. from methylcylohexane*

The molar selectivities of the products as a function of the methylcyclohexane conversion are shown in Fig.2. The isomerisation products are obtained with nearly 100% selectively up to a methylcyclohexane conversion of 50%. The maximum selectivity toward ring opening products is approximately 20% for a MC conversion of 70%. Above 80% MC conversion the selectivity toward cracked products is significant.

The distribution of the cylcoalkane isomerisation products is shown in Fig.3. The dimethyl-/ethyl- cyclopentane ratio increases up to a MC conversion of 40% where it levels off at 86/14.

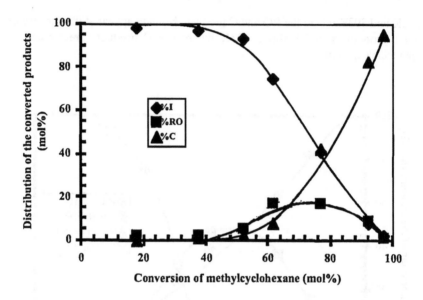

Figure 2 : Distribution of the converted products from methylcyclohexane

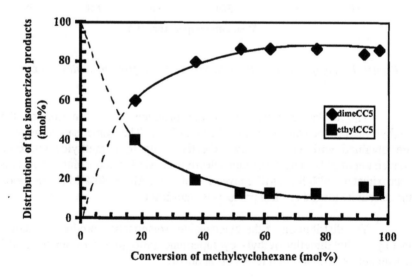

Figure 3 : Distribution of the converted products from methylcyclohexane

The Ring Opening product distribution is shown in figure 4. The data shown in Fig.2 indicates that the Ring Opening products (RO) are not appreciable below 60% methylcyclohexane (MC) conversion. On figure 4, we can note that at this level of MC conversion, the RO distribution attains a stable value i.e. 17% n-heptane (nC7), 65% methyl-hexane (meC6) and 17% dimethyl-pentane (dimeC5).

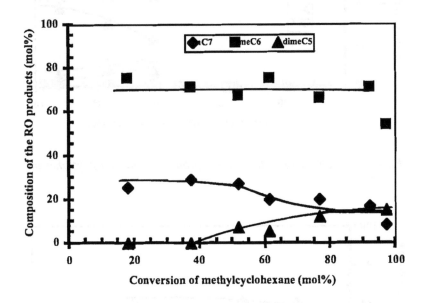

Figure 4 : *Distribution of the RO products from methylcyclohexane*

The molar distribution of cracked products (C) resulting from the hydrocracking of C-C bonds of the RO products is shown as a function of the MC conversion, Fig.5. The cracked products do not appear below 60% MC conversion. At higher conversions, the fraction of C4 to total cracked products is constant at about 50% and the propane fraction increases at the expense of the C5 and C6 alkanes. Methane and ethane are formed in negligible quantities.

Influence of the contact time : methylcyclohexane conversion, at 300°C, increases with increasing contact time up to a value of about 60% at 1h, see Fig.6.

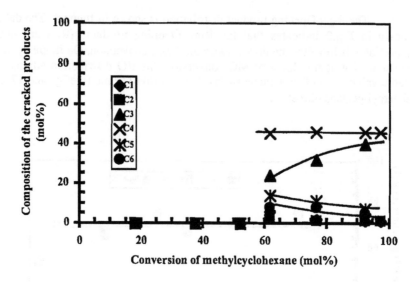

Figure 5 : *Distribution of the cracked products from methylcyclohexane*

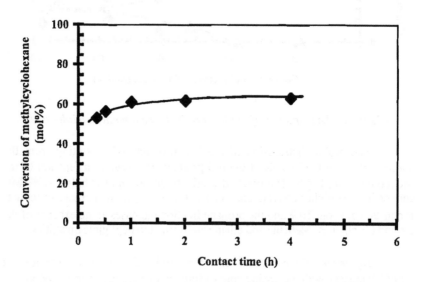

Figure 6 : *Conversion of methylcyclohexane vs contact time at 573K*

The products selectivities are reported in Fig.7. The yield of isomerisation products decreases with increasing contact time to probably be negligible at high contact time; whereas the yield toward cracked products increases and probably reaches 100% at high contact time. The ring opened product (RO) yield appears to reach a maximum of 18% at 3h contact time.

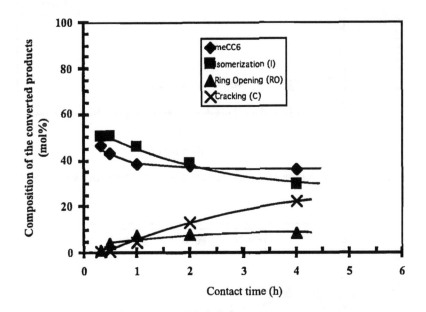

Figure 7 : Distribution of the effluents vs contact time from methylcyclohexane at 573K

The data shown in Fig. 8 indicates that dimethyl-/ethyl- cyclopentane ratio increases up to a maximum of 88/12 at 1h contact time.

The distribution of RO and C products are shown in Fig. 9 and 10 respectively; the same observations are made with figures 4 and 5.

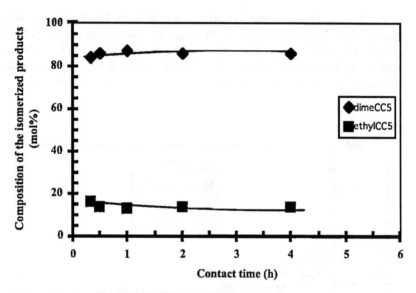

Figure 8 : *Distribution of the isomerized products vs contact time at 573K*

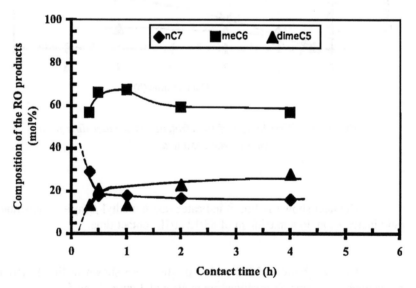

Figure 9 : *Distribution of the Ring Opening products (RO) vs contact time
from methylcyclohexane at 573K*

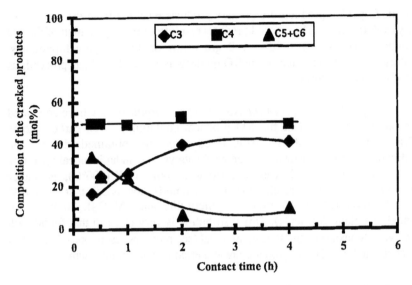

Figure 10 : *Distribution of the Cracked products (C) vs contact time from methylcyclohexane at 573K*

DISCUSSION

An investigation of the influence of the temperature and the contact time on methylcyclohexane isomerisation and ring opening on a bifunctional catalyst has enabled us to refine the reaction scheme.

MeCC6 conversion to alkylCC5 : the first step involved in the tranformation of MeCC6 into non-cyclic alkanes is its isomerisation into dimethyl- and ethylcyclopentanes. These isomers are the only products of the reaction up to a MC conversion of about 50% (see scheme 1). The products are at thermodynamic equilibrium above 553K (280°C) where the ratio of dimethyl- to ethylcyclohexane varies only slightly.

Opening of alkylcyclopentanes : the ring opening and cracking products (RO and C) are formed by secondary reactions as they do not appear before 50% conversion of methylcyclohexane (see Fig. 2). A detailed examination of Fig. 7 indicates that the ring opening products (RO) appear before the cracked products.

456   Mignard, Caillette and Marchal

The ring opening products are primary products resulting from the reaction of alkylcyclopentanes. The cracked products result from the degradation of the RO products and not from direct degradation of the alkylcyclopentanes. Unfortunately, degradation of the RO products to cracked products proceeds as rapidly as they are formed.

Equilibrium between the isomers of RO products : we have compared the molar distribution of the RO products containing seven carbon atoms (Fig. 4) with that of the isomerized products obtained during the hydroconversion of n-heptane on identical catalyst and under identical operating conditions (Fig. 11). In the case of n-heptane, above 573K (300°C), the molar distribution of products resulting from the isomerization, is constant : 20 mol% of n-heptane (nC7), 55 mol% of methylhexane (MeC6), 20 mol% of dimethylpentane (diMeC5). This distribution is very close to that found with methylcyclohexane as the starting material (see previous discussion). It can thus be concluded that the RO products are at thermodynamic equilibrium.

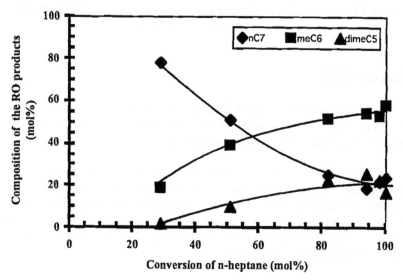

Figure 11 : Distribution of the isomerized products (I) from nC7

Cracked products : it has been observed that the percentage of cracked products as butanes (C_4) remains constant with increasing temperature.

Whereas, the fraction of cracked products as propane (C_3) increases and the fractions of pentanes (C_5) and hexanes (C_6) decreases. It is important to note that the quantities of methane (C_1) and ethane (C_2) are small with respect to those of C_5 and C_6. Thus, we can conclude that reactions 1a and 1b are minor reactions, eq. (1). Furthermore, we have carried out experiments using nC_7 and have found equimolar amounts of C_3 and C_4 with very low amounts of C_1, C_2, C_5 and C_6.

$$
\text{alkylCC}_5 \text{ or } nC_7 \xrightarrow{\quad\text{major}\quad} C_3+C_4 \qquad (1a)
$$
$$
\xrightarrow{\quad\text{minor}\quad} C_2+C_5 \qquad (1b)
$$
$$
\xrightarrow{\quad\text{minor}\quad} C_1+C_6 \qquad (1c)
$$

The major products resulting from nC_{10} cracking are C_4, C_5 and C_6 with very little production of C_1, C_2 and C_3 [6], eq. (2). Thus, the cracking of a C_{10} molecule could account for the abnormally high production of C_4 and C_5 hydrocarbon in our reaction. However, C_{10} hydrocarbon have not been detected in our experiments using classical analytical techniques. It is also clear that nC_7 does not react with C_3 to produce C_{10}, eq. (3), since previous experiments with nC_7 did not produce C_5 and C_6.

$$
nC_{10} \longrightarrow C_4+C_5+C_6 \qquad (2)
$$
$$
nC_7+C_3 \longrightarrow C_{10} \qquad (3)
$$

It is believed that the abnormally high C_5 and C_6 concentrations result from the hydrocracking of cyclic-C_{10} species which are produced from the reaction of a cycloheptane carbocation or olefin species and propylene or propylcation on the catalyst surface, eq. 4 :

$$
[CC_7]^+ + C^=_3 \text{ or } CC^=_7 + [C_3]^+ \longrightarrow [CC_{10}]^+ \qquad (4)
$$

The resulting transitory cyclic-C_{10} carbocation then immediately undergoes ring opening and hydrocracking reactions to produce C_4, C_5 and C_6 (eq. 2). As the temperature increases, the contribution of this secondary reaction involving a cyclic-C_{10} carbocation decreases. Our results have allow us to elucidate, step by step, the hydroconversion of methylcyclohexane to non-cyclic hydrocarbons.

CONCLUSION

Hydroconversion of methylcyclohexane has been studied with a bifunctional catalyst in hydroconversion operating conditions. The catalyst which has been used is a noble metal (Pt) deposited on a dealuminated HY zeolite. In our case, the apparent reaction scheme has been deduced :

$$MeCC_6 \longrightarrow \begin{pmatrix} EtCC_5 \\ \downarrow\uparrow \\ diMeCC_5 \end{pmatrix} \longrightarrow \begin{pmatrix} nC_7 \\ \downarrow\uparrow \\ MeC_6 \\ \downarrow\uparrow \\ diMeC_5 \\ \downarrow\uparrow \\ triMeC_4 \end{pmatrix} \longrightarrow C_3 + C_4$$

$$\downarrow +diMeCC_5$$
$$C\dot{C}_{10}$$
$$\downarrow \text{ very fast}$$
$$R\dot{O}_{10}$$
$$\downarrow \text{ very fast}$$
$$C_4 + 2C_5 + C_6$$

REFERENCES

[1]   M.F. ALVAREZ, F. RIBEIRO, F. RAMOA and M. GUISNET
        React. Kinet. Catal. Lett. (1987) 41 (2), 309-14
[2]   C. HENRIQUES, P. DUFRESNE, C. MARCILLY, and F. RAMOA-
        RIBEIRO Appl. Catal. (1986) 21 (1), 169-77
[3]   C. HENRIQUES, P. DUFRESNE, C. MARCILLY, and F. RAMOA-
        RIBEIRO Acta. Phys. Chem. (1985) 31 (-2), 477-86
[4]   G. GIANETTO, G. PEROT and M. GUISNET
        Actas Simp. Iberoam. Catal. (1984) 9th, vol. 2, 1661-2
[5]   J.A. MARTENS, M. TIELEN and P.A. JACOBS
        Catalysis Today 1 (1987) 435
[6]   J. WEITKAMP
        Ind. Eng. Chem. Prod. Res Dev. 1982, 21, 550-558

Author Index

Subject Index

Rare earth, 13
Reactor
  ebulated-bed, 21, 55, 61, 67, 219, 407
  fixed-bed, 21, 206, 236, 255, 281,
           330, 407
  hydropyrolysis, 91
  riser, 405
  slurry-phase, 143, 144
  stirred tank, 37, 72, 128, 177, 220,
           423, 435
Residue
  Alaska North Sloe, 57
  Amna, 147
  Arabian Light, 12, 57, 90, 115, 154
  capacity, 57
  Cold Lake, 36, 148, 221, 236
  hydrocracking, 39, 63, 65, 128
  hydropyrolysis, 101, 105
  hydrotreating, 58, 120
  hydrovisbreaking, 120
  Kuwait, 159, 207
  Lloydminster, 36
  Louisiana Sweet, 157
  Mexican Maya, 148
  Nigeria Medium, 147
  Shale oil, 71
    Venezuelan Boscan, 147
  Ring-opening, 448

Sediment, 47, 57, 63, 65, 119, 128,
  140, 163
Silica, 12, 181, 185
SIMS, 182
Smoke point, 236, 250, 301
Sodium, 12, 181, 185, 194
Solvent, 144
Sulfiding
  DMDS, 36, 223, 358, 359
  gas oil, 161, 207
  hydrogen sulfide, 2, 72, 358, 423
  kerosene, 237

Tar sand, 147, 175, 235
tert-butylamine, 256
Tetrahydrothiophene, 360, 366
Tetralin, 255, 259, 263, 424
Thiophene, 358, 361, 411
Toluene, 255

Vanadium, 7, 8, 11, 12, 181, 185,
  189, 195
Viscosity, 116

Zeolite, 128, 130, 133, 423